Cambridge checkpoint

CW01572694

Lower Secondary
Mathematics
TEACHER'S GUIDE

7

with Boost Subscription

Ric Pimentel
Frankie Pimentel
Terry Wall

Boost

HODDER
EDUCATION
AN HACHETTE UK COMPANY

The Publishers would like to thank the following for permission to reproduce copyright material.

Acknowledgements

Cambridge International copyright material in this publication is reproduced under licence and remains the intellectual property of Cambridge Assessment International Education.

End of section test questions and sample answers have been written by the authors. In assessment, the way marks are awarded may be different. References to assessment and/or assessment preparation are the publisher's interpretation of the syllabus requirements and may not fully reflect the approach of Cambridge Assessment International Education.

Third-party websites and resources referred to in this publication have not been endorsed by Cambridge Assessment International Education.

Every effort has been made to trace all copyright holders, but if any have been inadvertently overlooked, the Publishers will be pleased to make the necessary arrangements at the first opportunity.

Although every effort has been made to ensure that website addresses are correct at time of going to press, Hodder Education cannot be held responsible for the content of any website mentioned in this book. It is sometimes possible to find a relocated web page by typing in the address of the home page for a website in the URL window of your browser.

Hachette UK's policy is to use papers that are natural, renewable and recyclable products and made from wood grown in well-managed forests and other controlled sources. The logging and manufacturing processes are expected to conform to the environmental regulations of the country of origin.

Orders: please contact Hachette UK Distribution, Hely Hutchinson Centre, Milton Road, Didcot, Oxfordshire, OX11 7HH. Telephone: +44 (0)1235 827827. Email education@hachette.co.uk Lines are open from 9 a.m. to 5 p.m., Monday to Friday. You can also order through our website: www.hoddereducation.com

ISBN: 978 1 398 30072 9

© Ric Pimentel, Frankie Pimentel and Terry Wall

First published in 2011
This edition published in 2021 by
Hodder Education,
An Hachette UK Company
Carmelite House
50 Victoria Embankment
London EC4Y 0DZ

www.hoddereducation.com

Impression number   10 9 8 7 6 5 4 3 2 1

Year   2025 2024 2023 2022 2021

All rights reserved. Apart from any use permitted under UK copyright law, no part of this publication may be reproduced or transmitted in any form or by any means, electronic or mechanical, including photocopying and recording, or held within any information storage and retrieval system, without permission in writing from the publisher or under licence from the Copyright Licensing Agency Limited. Further details of such licences (for reprographic reproduction) may be obtained from the Copyright Licensing Agency Limited, www.cla.co.uk

Illustrations by Integra Software Services Pvt. Ltd., Pondicherry, India

Typeset in FSAlbert 12/14 by Integra Software Services Pvt. Ltd., Pondicherry, India

Printed in the UK

A catalogue record for this title is available from the British Library.

Contents

Contents

Section 3

Introduction to the *Teacher's Guide*

This series of books has been written specifically for students in schools following the revised Cambridge Lower Secondary Mathematics curriculum framework (0862) from 2020. The authors are experienced classroom teachers of mathematics at secondary and sixth-form level and have also worked in the UK and internationally.

The general aims of the course

The content of the *Cambridge Lower Secondary Mathematics Student's Book 7* is supported by the content in this book and online to provide a course which matches the Cambridge Lower Secondary curriculum framework at Stage 7 and builds the foundation for IGCSE work. As well as providing materials tailored to the specific content requirements of the Cambridge Lower Secondary Mathematics curriculum framework, the books in the series and the online resources also set out to increase the mathematics capital of students. This means extending their knowledge of maths and further developing a positive attitude to its study.

This series has been changed significantly in order to incorporate the aims and objectives of the new framework. In particular great emphasis has been placed on developing material for the new area of Thinking and Working Mathematically (TWM). It is not enough to learn and apply rules which enable students to come up with 'the answer'. Mathematics should be seen as a language which expresses complex ideas simply. In the introduction to the *Student's Book*, this idea is set out at the start.

Teachers might want to explain this idea in more detail at the start of the course and repeat it at regular intervals. Some of the points that teachers might want to make include:

- Mathematics is a language. The number system we use developed over hundreds of years.
- Mathematics is universal. In countries where the language and even the alphabet is different, mathematicians can 'talk' to each other.
- Mathematics simplifies language and is exact. For example, the sentence 'What do you get if you add seventeen and twenty-three together and take the total away from one hundred?' becomes $100 - 17 - 23 =$, or better $100 - (17 + 23) =$.
- It is said that 'a picture is worth a thousand words'. This is very true of mathematics. The sentence 'If a triangle has two angles of 45 degrees and 30 degrees, what is the size of the third angle, and what type of triangle is it?' can be shown very clearly in a sketch.

Beginning lessons with some short mental activity helps both to create a class atmosphere and to get the students thinking in a mathematical way. When students are encouraged to work in pairs for a few minutes, it has the added benefit for students whose first language is not English to get them talking in English and in 'Maths'. We have added some mental tests, with suggestions of how they might be used, to this *Teacher's Guide*.

In addition, teachers in many countries, who had used our *Cambridge Lower Secondary Mathematics* series, *IGCSE Mathematics* and *Core Mathematics for IGCSE*, were consulted and asked what they needed to help them to teach the new curriculum framework successfully. We have incorporated their suggestions into this new series of books.

The major points on which most teachers agreed were:

- That each book in the series should cover all the content as outlined in the Cambridge Lower Secondary Mathematics curriculum framework.
- That each book should provide plenty of opportunities for students to experience Thinking and Working Mathematically type questions at varying levels of difficulty.
- That each *Student's Book* should be set out in a linear fashion, so that it would be possible for a teacher to start at the beginning of the book if needed and then work through it in the order given.
- That there should be further exercises in a separate *Workbook* which could be used as homework or further class work. This *Workbook* would then provide a record of progress for the benefit of the teacher, act as a revision guide for students, and allow parents to see easily how their child was progressing.
- That additional assessments should be provided in the *Teacher's Guide* so that teachers could assess students at the end of each section.
- That detailed explanations should be included to help inform teachers as to what constituted a TWM type question and how they differ from traditional questions.

The separate 'write-in' *Workbook*, which accompanies each book in the series, provides further exercises that can be set for homework or further class work. These mirror the content and style of the questions in the *Student's Book*.

The *Teacher's Guide* and *Online Teacher's Guide* also contain additional assessment material, intended for both formative and summative purposes.

It is important that teachers refer to the detailed Cambridge curriculum framework and progression grids from Cambridge International on a regular basis so that they are confident that what is being taught matches the curriculum framework. The progression grid also shows what has been taught at each level and has examples.

This Cambridge Lower Secondary Mathematics series includes all of the content from the curriculum framework. Teachers can be confident that, taken together, the *Student's Book*, *Workbook* and this *Teacher's Guide* will provide full support to deliver the programme.

In this *Teacher's Guide*, we are taking each unit of the *Student's Book* and outlining for teachers:

a the stage 7 strands
b the prior knowledge that students should already have
c comments and suggestions which will help in lesson presentation.

We refer often to displays, preferably made by students. These are not only fun to make, but are constant forms of revision. Students will always look at their own displays.

This *Teacher's Guide* includes teaching tips and advice for teaching maths to learners whose first language is not English. Online you will find plenty of ideas for helping learners develop maths vocabulary in English as well as reading and listening to maths word problems in English. These include both strategies for teachers and students and techniques to help boost learner confidence in comprehending maths word problems. Additionally, there are ten language development activities. You can use these as flexible models that can accommodate the content you are teaching. Most of these activities are designed to familiarise students with techniques for developing maths vocabulary, taking notes, organising ideas and comprehending word problems. They can be used either in class, for example during self-study sessions, or assigned as homework.

Features of the *Student's Book*

To try to meet all of the objectives set for us by teachers we have set out the *Student's Book* in three clear sections. Each section has the following characteristics:

- It starts with a historical note so that students can begin to engage with the rich cultural history of mathematics.
- It contains ten units, covering topics to do with Number, Algebra, Geometry and measures and Statistics and probability. These do not follow a set pattern but are organised so that students cover work topics needed later in the book.
- Each unit starts with the curriculum content covered in it followed by a detailed explanation of the topic including worked examples and exercises.
- The questions within each exercise are graded according to difficulty (green, amber and red). It is expected that most students will be able to complete the green questions and only a few of the more confident students be able to complete the red questions.
- Some of the questions will have symbols beside them to help you answer the questions. Look out for these symbols throughout the *Student Book*:

  - ⭐ A question which specifically addresses the TWM content of the framework is also highlighted using a star. See more about TWM on page x.

  - 🖩 This indicates where you will see how to use a calculator to solve a problem.

  - ✖ These questions should be answered without a calculator.

  - 🔗 This tells you that content is related to another subject.

  - 🔊 This tells you that content is available as audio. All audio is available to download for free from www.hoddereducation.com/cambridgeextras

  - 🖱 There is a link to digital content at the end of each chapter if you are using the Boost eBook.

Each unit in the *Student's Book* has most or all of the following features:

- Learning objectives for the unit.
- Let's talk boxes – talk with a partner or a small group to decide your answer when you see this box.
- Worked examples – these show how to approach answering a question.
- Key information – these give hints or pointers to solving a problem or understanding a concept.
- Thought bubbles – these highlight ideas and things to think about.
- Exercises – there are lots of activities to help students learn. These are assigned a level according to difficulty. Green are introductory, amber are more challenging and red are challenging questions.
- End of section review questions – these feature several areas covered in the section that can be used when the teaching of the section is completed or revisited later as part of a revision programme.
- Glossary words – words which are printed in red in the text are featured with a definition in the glossary at the end of the *Student's Book* to develop the student's mathematical vocabulary further.

Features of the *Workbook*

Each *Student's Book* has an associated *Workbook*. The units in the *Workbook* are set out in the same sequence as the units in the *Student's Book*. There are headings in each unit to help you relate them to the sections in the *Student's Book* so that you can integrate your work.

The *Workbook* provides extra questions. Space is provided for the students to write their answers. The answers are included in this book.

Features of the *Teacher's Guide*

Each unit in the *Student's Book* is supported by a unit in this book which is integrated with support in the *Online Teacher's Guide*. The units of the textbook are set out so that they form a detailed lesson plan for teachers with explanation and worked examples. They may be of particular help to teachers who are not maths specialists and/or are not fluent in mathematical terms.

Prior knowledge

It is important that teachers are aware of the work done by students in previous stages, as well as earlier in the current course. Many students will have followed the Cambridge framework curriculum, but even those who have not will need to have covered the work necessary to progress. The Prior knowledge sections in this *Teacher's Guide* summarise what students will have experienced before in their learning journey, including in earlier stages and within Stage 7.

Teachers can refer to the Cambridge curriculum framework which will give more detail and examples of what students have learned in earlier stages and may find it useful to look ahead to see how students will progress at Stage 8.

Displays of students' work is referred to often in the Prior knowledge section of this guide. Displaying students' work is important both as a help with revision and to 'set up' a classroom as a maths class. Students like to see their work on the wall of a room and have a constant reminder of mathematical ideas.

Learning objectives

Each unit has a table of learning objectives which shows where in the unit the objectives are addressed.

Background information

The summary provides information about the topics covered in the unit.

Terminology

This section features the mathematical terms printed in bold in the *Student's Book*. It can be used in vocabulary work involving the glossary at the end of the *Student's Book* and in conjunction with the ESL support materials in the online Boost section of the *Teacher's Guide*.

Lesson ideas

These are suggestions on how you may group the information and activities to break up the unit into a series of lessons. They are based on the activities in the *Student's Book* but you may also like to add some enrichment activities to some lessons or even make one or more extra lessons based on them.

Starter activities

At the start of each unit in the *Teacher's Guide* there is a Starter activity. These provide ideas for how to introduce each unit or topic. Worksheets for starter activities are available as printable files; to access the online teaching resources that accompany this *Teacher's Guide* please visit boost-learning.com to download the printable starter worksheet.

TWM activity notes

For Thinking and Working Mathematically exercises in the *Student's Book* there are activity notes to further explain the activity.

Support and extension activities

These activities provide support and further practice for students who are struggling, or additional stretch activities for those who found the starter activity straightforward. Worksheets for support and extension activities are available as printable files; to access the online teaching resources that accompany this *Teacher's Guide* please visit boost-learning.com to download the printable worksheets.

Answers

Answers to all the questions in the *Student's Book* and the *Workbook* are given in this book and separately collated in the *Online Teacher's Guide.* In some instances, a student may wish to display extra knowledge to a straightforward question, or the question may encourage the student to think more widely. The answers in this book give the simple response to straightforward questions but there may be other responses that can be added to these that take into account the students' previous experience. These extra responses may be used to give the students extra credit for their work or provide additional material that may be used to assess the students' attitude, aptitude and general progress.

The *Online Teacher's Guide*

The *Online Teacher's Guide* provides teacher support in conjunction with this *Teacher's Guide.* Both the book and online support should be used together to provide your students with the mathematical knowledge and understanding to meet the requirements of the Cambridge Lower Secondary Mathematics curriculum framework.

To access the online teaching resources that accompany this *Teacher's Guide* please visit boost-learning.com. All starter activities and other activities in this *Teacher's Guide* are also available as printable worksheets on Boost.

Online content

The contents of the *Online Teacher's Guide* are listed below:

1 eBook of the *Teacher's Guide*
2 Online introduction
3 Starter activity printable worksheets
4 Support and other activities printable worksheets
5 *Student's Book* and *Workbook* exercise answers
6 Mental strategies guide and supplementary mental tests and answers
7 End of section tests and answers for the three sections of the *Student's Book*
8 Knowledge tests for each unit of the *Student's Book*
9 Supplementary worksheets and answers
10 Glossary and bilingual glossary
11 ESL support documents
12 Flashcards and answers

Mental strategies

Calculating mentally is an important skill for students to have as it requires deep conceptual understanding, number reasoning and number sense.

The mental strategies' learning objectives have been removed from the framework; however, this was to avoid it being taught as a bolt-on topic. Exploring mental strategies should be an integral part of most topics in order for students to develop a range of different strategies which can be applied in different contexts.

Although the *Student's Book* does occasionally highlight exercises where a calculator should not be used, students should not automatically reach for a calculator in those exercises where the symbol is not used. Instead students should be encouraged to use mental strategies as much as possible.

Notes on mental strategies are included, together with a number of mental arithmetic tests, which emphasise both the importance of mental working and its central importance to all aspects of mathematical thinking.

Calculator strategies

The calculator, when used properly, is a valuable tool to aid students with their understanding of number. Throughout these books students are encouraged to use calculators to this aim. In particular, students should develop a sense of when it is best to use calculators and when mental and written strategies are better.

As part of calculator use, students should be taught how to use a calculator efficiently and to become familiar with its many functions. This series of books introduces, when appropriate, some of the main calculator functions and the calculator keys to use.

Thinking and Working Mathematically

Thinking and Working Mathematically (TWM) is an important approach to mathematical thinking and learning that has been incorporated throughout this book.

Questions involving TWM differ from the more straightforward traditional question-and-answer style of mathematical learning in that their aim is to encourage students to think more deeply about the problem involved, make connections between different areas of mathematics and articulate their thinking.

Cambridge International has identified the eight characteristics of TWM as:

Characteristic	Definition
Specialising	Choosing an example and checking if it satisfies or does not satisfy specific mathematical criteria.
Generalising	Recognising an underlying pattern by identifying many examples that satisfy the same mathematical criteria.
Conjecturing	Forming mathematical questions or ideas.
Convincing	Presenting evidence to justify or challenge a mathematical idea or solution.
Characterising	Identifying and describing the mathematical properties of an object.
Classifying	Organising objects into groups according to their mathematical properties.
Critiquing	Comparing and evaluating mathematical ideas, representations or solutions to identify advantages and disadvantages.
Improving	Refining mathematical ideas or representations to develop a more effective approach or solution.

The TWM questions appear alongside other question types within the exercises of the book. They are flagged up with this icon 🌟 and, in this way, students will be aware that the question is likely to demand a higher level of mathematical understanding and communication.

It must be stressed that, often with TWM style questions, there is no one 'correct' way of approaching the question. Students may approach any given question in a myriad of different ways and this in itself provides an excellent opportunity for the teacher to engage in discussion with the students over their thinking and strategies and, if appropriate, get them to reflect on possible alternatives to their own method.

In order to highlight the characteristics of TWM questions, a selection of ten examples have been chosen from the *Student's Book* and explained in detail within the relevant unit of this *Teacher's Guide*. The table below maps these ten examples.

Unit	Exercise	Question	Context	*Teacher's Guide* page number	*Student's Book* page number
1	1.3	3	Students have been introduced to multiplication and division by integers including negative numbers.	3	7
4	4.1	3	Students have just covered how to calculate the area of a triangle using the formula and have been told of the need for the height to be measured at right angles to the base.	20	21
10	10.1	3	Students have covered how to calculate the theoretical probability of equally likely events and how this probability can be expressed as either a fraction, decimal or percentage.	46	72
12	12.1	2	Students know how to calculate the mean, median, mode and range for a list of numbers and have been introduced to the idea that depending on the context, one value may be more useful than another.	56	90
15	15.5	1	Students will already have covered how to multiply decimals by powers of 10, in addition to the use of mental strategies for simplifying calculations. Calculators are discouraged for this question.	72	128
17	17.4	4	Students have covered the sum of angles around a point equals 360° and have been introduced to the properties of tessellating shapes.	83	148
18	18.1	9	Students have encountered how to write algebraic expressions and how to substitute values into it.	89	155
25	25.3	1	Students will have covered coordinates in all four quadrants as well as the properties of quadrilaterals.	116	203
26	26.1	4	Students have been introduced to square numbers and square roots.	121	208
30	30.1	2	Students have been introduced to real-life graphs in the context of travel graphs involving distance and time. The significance of a straight-line distance–time graph has also been introduced.	136	239

Questioning strategies and evaluation of student responses

This section will look at three TWM questions in detail and highlight opportunities for teachers to develop TWM through their questioning. The questions should elicit student responses which in themselves will be the indicators of TWM. Each question also highlights which TWM strand it is likely to cover, although that will of course be dependent on how the student responds to the question.

These are not meant to be exhaustive. Different students respond in different ways. Teachers too will have their own ways of questioning, suited to the individual needs of the student, which may differ from those given here.

Generally, it is better to avoid giving the actual answer to a question: instead try to pose a further question to enable the student to answer their own problem. Even if the student is simply seeking confirmation to an answer they have written down, asking them to 'explain how they got to their answer' or 'why they think it's correct' will be developing their ability to present a convincing argument as well as highlighting their reasoning.

A teacher cannot, though, be in all places at once, so getting students to explain their ideas, reasoning and conclusions to other students is an excellent way of developing their TWM skills.

Exercise 15.5

Context:

Students will already have covered how to multiply decimals by powers of 10, in addition to the use of mental strategies for simplifying calculations. Calculators are discouraged for this question.

1 a The rectangle below has the dimensions shown.

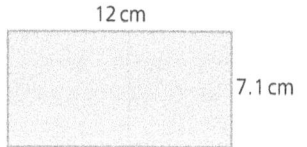

12 cm

7.1 cm

> How would you normally work out the area of a rectangle? Why is this trickier? (TWM skill: characterising)

> What multiplications could you do without a calculator that would be an estimate for the area? (TWM skill: specialising)

Explain why the area of the rectangle can be calculated by the calculation

$(7.1 \times 10) + (7.1 \times 2)$

> Are there different ways of splitting this rectangle? Describe them to me. (TWM skills: convincing, characterising)
>
> So far we have added areas together to work out the multiplication. How would the area be calculated by subtraction? (TWM skills: specialising, generalising, convincing, characterising, improving)
>
> Which method is easier? Does it change depending on numbers you are working with? (TWM skills: generalising, convincing, characterising, classifying, critiquing)

b Work out the area of the rectangle below.

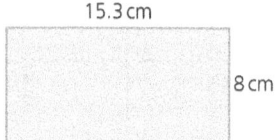

15.3 cm

8 cm

> Are there different ways of splitting this rectangle? Describe them to me. (TWM skill: convincing, characterising)
>
> So far we have added areas together to work out the multiplication. How can the area be calculated by subtraction? (TWM skills: specialising, generalising, convincing, characterising, improving)
>
> Which method is easier? Does it change depending on numbers you are working with? (TWM skills: generalising, convincing, characterising, classifying, critiquing)

Exercise 18.1

Context:

Students have encountered how to write algebraic expressions and how to substitute values into it.

9 There are three jigsaw puzzle boxes. The first box has n pieces, the second has $3n$ pieces and the third has $2n + 40$ pieces.

> Explain what you think is meant by n. (TWM skill: convincing)

a Is it possible to tell which box has the least number of pieces? Justify your answer.

> Is there only one possible box that must have the smallest number of pieces? How do you know? (TWM skills: generalising, characterising, classifying)
>
> Could you label a box which would have less than the one with the least here? How do you know it would have less? (TWM skills: generalising, conjecturing, convincing, characterising)

b Is it possible to tell which box has the most number of pieces? Justify your answer.

> Is there only one possible box that must have the most number of pieces? How do you know? (TWM skills: generalising, characterising, classifying)
>
> Should we just try random values for n and see what happens? Is there a quicker way? (TWM skills: specialising, conjecturing, convincing, critiquing)
>
> What characteristics determine whether there is only one answer of several possible answers? (TWM skills: generalising, convincing, characterising, classifying)
>
> Are there critical values for n which change the order of which boxes have the least/most number of pieces? Is there a quick way of finding them? (TWM skills: generalising, convincing, critiquing, improving)

Exercise 26.1

Context:

Students have been introduced to square numbers and square roots.

4 A tiler has 100 square tiles for tiling two square panels in a bathroom.

100 tiles

> What type of number is 100? What properties does it have? (TWM skills: specialising, convincing, characterising)

a Explain why the panels cannot be of the same size.

> If they were the same size, what's the largest they could be? How many tiles would be left over? (TWM skills: specialising, convincing, characterising)
>
> If boxes come in multiples of 100, is there a box which will produce two squares of the same size? Is there a pattern to the boxes with this property? (TWM skills: specialising, generalising, conjecturing, convincing, characterising, classifying)

b Each tile is 10 × 10 cm. What are the dimensions of the two square panels if all 100 tiles are used?

> Is there more than one possible solution? Prove your answer. (TWM skills: specialising, convincing, characterising)
>
> How many different combinations of square panels are possible if we still have to use a complete box of 100 tiles, but we are allowed to make any number of square panels and they can be the same size? (TWM skills: specialising, convincing, characterising, classifying)
>
> Describe how you found all the possible combinations. What methods did you use to make sure you didn't miss a combination out? Did you change your method whilst doing it? If so, how and why? (TWM skills: all)

Thinking and Working Mathematically question list

These tables identify Thinking and Working Mathematically questions in the *Student's Book, Workbook* and online Boost knowledge tests. The 'TWM skill focus' column indicates the main focus characteristic, but there may be other TWM characteristics addressed, depending on the approach to the question.

Student's Book

Exercise	Colour code Green	TWM skill focus	Colour code Amber	TWM skill focus	Colour code Red	TWM skill focus
1.1					5	generalising, convincing
					6	convincing
1.3			2	specialising	3	generalising convincing characterising
2.1	1	classifying	2, 3	characterising		
2.2	1	classifying			4	convincing
3.2			1	convincing		
			2, 3	critiquing		
3.3	1	convincing, improving	3	critiquing improving		
	2	critiquing, improving				
3.4			1	critiquing	2	critiquing
4.1			3	convincing, characterising	6	generalising
			4	specialising		
			5	generalising		
4.2					4	conjecturing
					5	convincing
5.1	2	convincing	5	convincing	6	specialising
6.1			5	convincing	6	convincing
6.4			2	specialising		
7.1			2	specialising		
			3	convincing		

Exercise	Colour code Green	TWM skill focus	Colour code Amber	TWM skill focus	Colour code Red	TWM skill focus
7.2			2	specialising		
			3	critiquing, classifying		
7.3	2	specialising characterising	3	conjecturing	5	
			4	specialising		
7.4			3	conjecturing		
			4	classifying		
7.5	1	improving	2	critiquing	3	classifying
8.1			3	convincing characterising	4	specialising convincing
8.2			6	specialising	7	conjecturing
8.3			4	generalising	5	generalising
					6	conjecturing
8.4	3	specialising	4	generalising	6	conjecturing
					7	convincing
9.1	2	generalising	4	specialising	5	convincing
9.2	2	specialising	3	generalising conjecturing		
10.1	2	generalising	3	specialising characterising classifying	5	conjecturing
11.1					7	specialising
11.2			2	conjecturing	5	conjecturing
11.3					7	convincing
11.4	3	convincing	4	conjecturing		
12.1	2	convincing, characterising, critiquing	4	generalising	9	generalising
			5	specialising	10	convincing
	3	convincing	6, 7	conjecturing		
			8	convincing		
12.2	1	specialising			3	conjecturing
13.1	1	classifying	3, 4, 5	specialising		
13.2			3	specialising	4	conjecturing
13.3					7	critiquing
13.4	1, 2, 3, 4	classifying	5, 6	classifying		
13.5					7	characterising
14.1			3	convincing	4	convincing
14.2	2	conjecturing	4	conjecturing	5	convincing
15.1			2	classifying	4	generalising
15.2			2, 3	conjecturing	4, 5	generalising
15.3	1	conjecturing	2	conjecturing	4	generalising
15.4	2	classifying	4	convincing	5	generalising

Exercise	Colour code Green	TWM skill focus	Colour code Amber	TWM skill focus	Colour code Red	TWM skill focus
15.5	1 2	convincing, characterising, critiquing, specialising conjecturing	4 5	generalising conjecturing	7	specialising
16.1	2	classifying generalising	3 4	convincing specialising	5	convincing
16.2	1, 2	critiquing	3, 4 5	convincing conjecturing	6	conjecturing
17.2					3	specialising
17.3	1, 2, 3	convincing	4, 5	convincing	6	convincing
17.4			3	specialising convincing characterising	4	specialising convincing
17.5	1	generalising				
17.6	1	generalising				
17.7					3, 4	convincing conjecturing
18.1			5, 8 6, 7	specialising conjecturing	9 10	specialising, conjecturing, convincing, characterising specialising, convincing
19.1	1 2	improving critiquing, improving	3, 4	critiquing improving	5	critiquing improving
20.3			2, 3	conjecturing	4 5	conjecturing convincing
21.1	1, 2	generalising	3	generalising	4	generalising
21.3			3	generalising convincing	4	generalising
22.1			7	conjecturing	10	generalising
22.2			7, 8	conjecturing	10, 11, 12	conjecturing
23.1	1	characterising	3	convincing	5	characterising
23.2					5	convincing
24.1			5	generalising	7	generalising
24.2					4	specialising critiquing
25.2	1, 2	characterising	4	conjecturing	5, 6	conjecturing
25.3	1	generalising convincing characterising	2	specialising	3	conjecturing

Exercise	Colour code Green	TWM skill focus	Colour code Amber	TWM skill focus	Colour code Red	TWM skill focus
26.1	2	conjecturing	4	specialising generalising conjecturing convincing classifying	5	specialising generalising conjecturing convincing characterising classifying
26.2	3	specialising generalising convincing characterising classifying			7	specialising generalising convincing characterising classifying
27.1	1	generalising	4	generalising	6	convincing
27.2			3	specialising classifying	5	convincing
					6	conjecturing
28.2					16	conjecturing
28.3					3	conjecturing
28.4					7	conjecturing
					8	convincing
29.2			5, 6, 7	generalising	8, 9	generalising
29.3					5	convincing
29.4			6	conjecturing		
30.1	1	specialising, classifying	3	specialising convincing characterising classifying critiquing	9	conjecturing
	2	specialising, convincing, characterising, classifying, critiquing				
30.2			2	convincing	4	specialising classifying
			3	conjecturing		

Workbook

Exercise	Colour code Green	TWM skill focus	Colour code Amber	TWM skill focus	Colour code Red	TWM skill focus
1.1			4, 5	conjecturing		
2.1–2.3			3	specialising	7	classifying
			4	classifying		
3.3			2	conjecturing		
4.1			3	conjecturing		
4.2					2	conjecturing
5.1	2	specialising				
6.1–6.2			3	conjecturing		
7.1–7.2					4	classifying
7.3–7.4	1	classifying	2	conjecturing		

Exercise	Colour code Green	TWM skill focus	Colour code Amber	TWM skill focus	Colour code Red	TWM skill focus
8.1–8.3			5, 6	generalising		
8.4			2	convincing		
9.1–9.2					6	generalising
					7	convincing
10.1			6	specialising		
11.1			4	convincing		
11.2			2	conjecturing	3	conjecturing
12.1–12.2			3	specialising	7	convincing
13.1–13.2			2, 3	specialising		
14.1–14.2	2	conjecturing	4	specialising	5	convincing
15.1			1, 2	classifying		
15.2			1	conjecturing	2, 3	specialising
15.3			1	conjecturing		
			2	generalising		
15.4					1	conjecturing
16.1–16.2			4, 5	convincing		
17.3–17.4			1	convincing		
17.5–17.7	1	convincing			2	convincing
18.1–18.2			3	critiquing	5	improving
20.3					2	conjecturing
20.4			3	conjecturing		
21.1	1	specialising			2	generalising
21.2					2	convincing
21.3			2	convincing	4	generalising
			3	generalising		
22.1			6	generalising	7	critiquing
22.2			3	conjecturing	4	conjecturing
23.1–23.2					3	specialising
24.1–24.2					6	convincing
25.1–25.2			3	specialising	4	specialising
25.3			2	characterising		
26.1–26.2			3	conjecturing	5	convincing
27.1–27.2					4	specialising
28.1–28.3			11	convincing		
28.4					5	conjecturing
29.1–29.2			6	generalising	7	conjecturing
29.3					4	convincing
29.4			3	convincing		
30.1–30.2			1	convincing		

Online knowledge tests

Knowledge test	Colour code Green	TWM skill focus	Colour code Amber	TWM skill focus	Colour code Red	TWM skill focus
1	3	classifying	5	classifying	6	convincing
					9	generalising
2	5	classifying	6	generalising		
3			2, 3	convincing	4, 5	critiquing
4			1, 3	specialising	6, 7, 8	conjecturing
5			3, 4, 5	convincing		
6			4	convincing	5	classifying
7			2	classifying	8	classifying
			5, 6, 7	conjecturing	9	conjecturing
8	2, 3	characterising	9, 10	conjecturing		
	4	conjecturing				
9			2, 7, 8, 9, 10	specialising classifying		
10	2	generalising	4	generalising	6, 7, 8	generalising
11			5	generalising	8	generalising
12			5	conjecturing		
13			4, 6, 7	classifying	8	conjecturing
14			4	conjecturing		
15			5, 7, 9, 10	classifying		
16	1	generalising			5	conjecturing
19			3, 4	critiquing	5	conjecturing
20					6	conjecturing
21			4, 7, 8	generalising	9, 10	generalising
22			7, 8, 9	conjecturing	10	conjecturing
23	2, 4	characterising	3, 5	characterising	6	characterising
25			5	characterising	8	conjecturing
26	1	classifying	5	classifying		
27			3, 4, 7	classifying	8	generalising
			6	generalising		
28			6	conjecturing	10	conjecturing
29			5, 6, 10	conjecturing	7	conjecturing
30			1, 2 3, 4,	classifying	6	classifying

Thinking and Working Mathematically

Plenty of research evidence shows that, despite previous attainment, learners who have developed the mathematical use of their natural powers can display an improved disposition towards mathematics. This is shown in improved performance and even answering exam questions they have not seen previously.

What does it mean to develop one's natural powers?

- It means taking opportunities to imagine a situation, to enter into it in order to become aware of underlying relationships, and it means expressing what is imagined in gesture, action, picture, diagram or symbols.
- It means that upon encountering some unfamiliar generality, the learner spontaneously tries some examples, not simply to get answers, but rather in order to detect underlying relationships which may apply in all cases (in other words, becoming independent learners).
- It means that upon encountering some specific situation, whether inside or outside the classroom, learners look for what can be varied without changing the outcome (for example, the answer to some task, or the method of tackling some problem).
- It means making conjectures and then not believing those conjectures; trying to justify them to yourself, to a friend and ultimately to a sceptic.
- It means characterising and classifying types of problems, so that you are more likely to recognise a problem to be 'of this type' in the future. It also means developing a rich concept image of mathematical concepts through appreciating the various features that characterise such an object, often through constructing your own examples.
- It means being critical of your own conjectures, as well as of other people's.
- It means refining your conjectures, including your personal narrative, so as to be more precise and more appropriate.

These natural powers are developed by setting tasks in which at least some learners use them spontaneously; then drawing attention to these actions. When another opportunity arises, referring back to experience, and where necessary directly prompting learners to act. Over time, the prompts become less and less direct, the learners more and more spontaneously taking the initiative. This is known as *scaffolding and fading*.

What is vital in promoting the use of learners' own powers is the choice of pedagogic action to initiate, whether it is beginning work on a new problem by imagining the situation, perhaps referring back to having done this also in the past, and how it is something they can do for themselves when studying, or whether it is prompting learners to try to articulate which actions were effective and promoting the construction of a personal story about a topic, including the concepts involved and the procedures which are used to resolve problems.

Tasks are simply tasks. Pedagogic actions involve the initiation of activity. To be effective, you need to have a positive relationship with learners so that they trust that what you set them can be achieved. Most importantly, it involves being aware yourself of the key ideas, concepts, procedures and pervasive mathematical themes involved in task or topic. Most particularly it means being aware of what at any moment *you* are attending to, so that you can direct learner attention appropriately.

If at the end of a year, learners require the same hand-holding support when meeting a new topic, procedure or problem that they did at the beginning of the year, their powers have not been developing. Thinking and Working Mathematically applies not simply to extension problems, or even to problems more widely, but to encountering new topics, new concepts or new procedures.

John Mason

SECTION 1

Addition, subtraction, multiplication and division

 The Section 1 introduction on very early mathematics and counting could relate to Cambridge Lower Secondary English as students are encouraged to engage with the rich cultural roots of mathematics through the introduction.

Prior knowledge

In previous learning, students will have had experience of adding, subtracting and estimating with integers including calculations where one integer is negative. Students will also have had experience of multiplying and dividing whole numbers up to 10 000 by another whole number of 1 or 2 digits.

Objective overview

Learning objective	Objective code	Student's Book pages	Workbook pages	Teacher's Guide pages	Online resources
Estimate, add and subtract integers recognising generalisations.	7Ni.01	2–7	2–4	1–6	Flashcards Unit 1
Estimate, multiply and divide integers including where one integer is negative.	7Ni.03	2–7	2–4	1–6	Knowledge test Unit 1

Background information

Being able to carry out calculations without the help of a calculator is an important skill. This unit looks at some methods for doing calculations and the importance of checking them.

Use of both mental and written skills is examined in this unit, including the use of number bonds, making estimations and spotting number patterns.

Students will learn to use all four operations (addition, subtraction, multiplication and division) with both positive and negative numbers.

Terminology

Students should recognise and use the word **integer** to mean a whole number (i.e. a number which is not fractional). Integers may be positive or negative.

Lesson ideas

Mental arithmetic can be a good way to revise and extend this topic. Firstly, by getting students to round numbers (say ten examples from the teacher), e.g. 53 rounds down to 50, 27 rounds up to 30. Then without getting an answer, ask how would 53×27 be estimated? (Ten similar questions from the teacher.) For example, 50×30. Finally, mentally estimate 69×21, answer 1400. (Ten questions written on the board.) Teachers should perhaps at the end of the lesson let students think of, and ask, each other similar questions. The rest of the lesson can be taken from the *Student's Book*.

Starter activity

Number cards

1 Arrange the five number cards so they show the i) greatest ii) least possible number.

| 2 | 6 | 9 | 1 | 5 |

2 Use the number cards so these calculations have the i) greatest ii) least possible answers.

a ☐ + ☐

b ☐ − ☐

c ☐ × ☐

d ☐ ☐ + ☐ ☐

e ☐ ☐ − ☐ ☐

f ☐ ☐ × ☐ ☐

g ☐ ☐ ☐ + ☐ ☐

h ☐ ☐ ☐ − ☐ ☐

i ☐ ☐ ☐ × ☐ ☐

Answers

1 i 96 521   **ii** 12 569

2 a i 9+6=15   **ii** 2+1=3

   b i 9−1=8   **ii** 2−1=1 or 6−5=1

   c i 9×6=54   **ii** 2×1=2

   d i 95+62=157 or 92+65=157   **ii** 15+26=41 or 16+25=41

   e i 96−12=84   **ii** 61−59=2

   f i 92×65=5980   **ii** 26×15=390

   g i 962+51=1013, 961+52=1013, 951+62 =1013 or 952+61=1013

     ii 126+59=185, 129+56 =185 or 159+26 =185

   h i 965−12=953   **ii** 125−96=29

   i i 952×61=58 072   **ii** 126×59=7434

TWM activity notes

Exercise 1.3 is explained here as an exemplar of Thinking and Working Mathematically (TWM), detailed in the Introduction to the *Teacher's Guide*, page xi.

Q3 $36 \times 18 = 648$

State the answer to each of the following calculations and explain how the above calculation was used to work it out.

a $648 \div 18$
b -36×18
c 18×18

d $-648 \div 9$
e -18×9
f $-1296 \div 36$

Although formal methods of multiplication and division are an important part of mathematics, so is the ability to spot patterns in order to simplify calculations. Here students are asked to *state* the answer of each calculation, implying that a formal calculation is not required. By looking at the given initial calculation and answer, students are asked to use this to state the answer to the subsequent calculations. This can be done through an understanding of the question's relationship to the original calculation. In addition, students are asked to explain their reasoning.

For example, if $36 \times 18 = 648$ then $18 \times 18 = 324$ as one of the numbers has been halved and so the answer must be halved too.

TWM characteristics: Generalising   Convincing   Characterising

This question if written in a standard way could have been presented as follows:

Q3   Work out the answer to the following calculations:

a $648 \div 18$
b -36×18
c 18×18

d $-648 \div 9$
e -18×9
f $-1296 \div 36$

Here students are asked to simply work out the answers to the calculations, something they would do possibly using formal methods of multiplication and division.

Support activity

Arithmagons

The number in each square is the sum of the numbers in the circles adjacent to it.

Fill in the missing numbers.

1
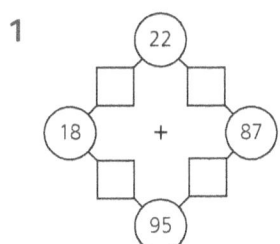

2
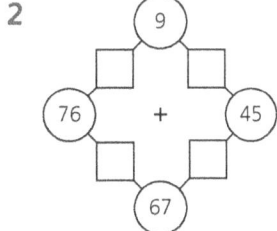

3
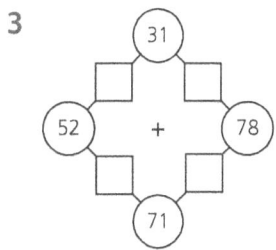

4
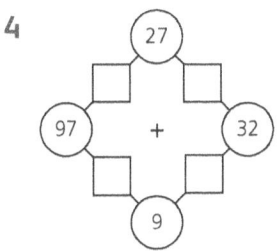

5
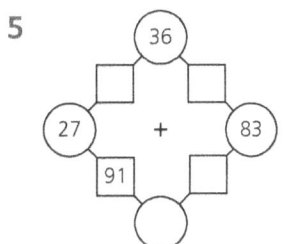

6
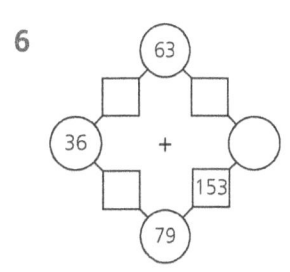

The number in each square is the product of the numbers in the circles adjacent to it.

Fill in the missing numbers.

7
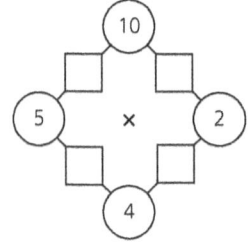

8
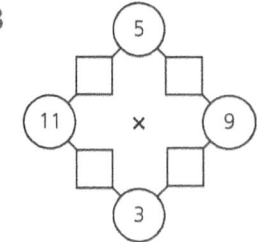

9
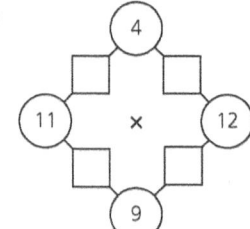

10
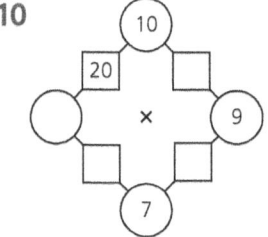

11
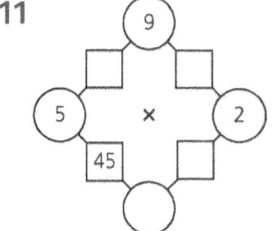

12
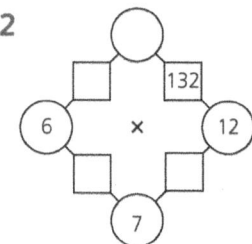

Answers

1
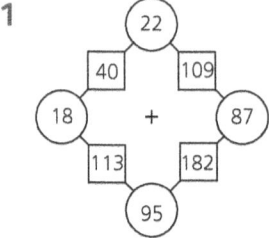

2
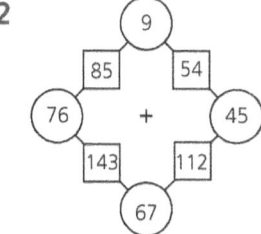

3
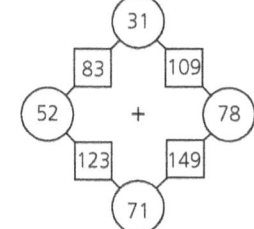

4
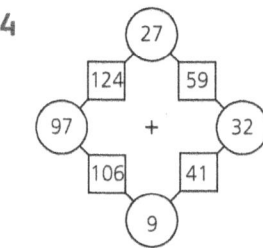

5
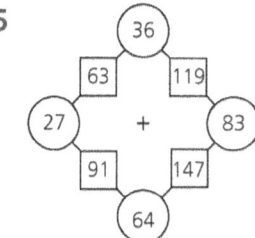

6
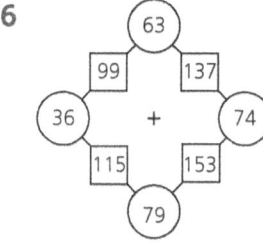

7
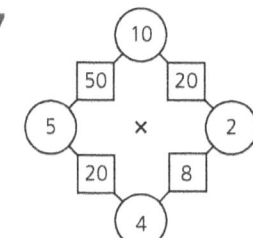

8
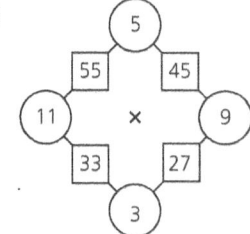

9
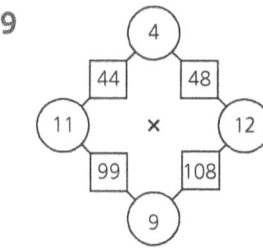

10
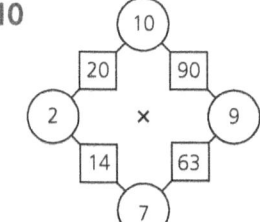

11
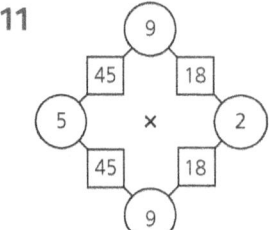

12
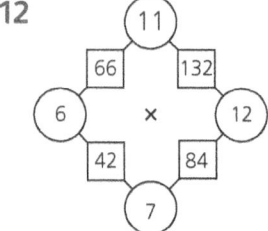

Student's Book answers

Exercise 1.1 (page 4)

1 a 50 b 70 c 100
   d 50 e 90 f 22
   g 11 h 33 i 6
   j 26

2 1710 km

3 a 150 b 390 c 220
   d 520 e 590 f 60
   g 210 h 320 i −20
   j 80

4 a 495 ml b 265 ml c 35 ml

5 a

2	9	4
7	5	3
6	1	8

   b i

52	59	54
57	55	53
56	51	58

   ii Each number is 50 more than the
       original number, therefore each row,
       column etc. which has three numbers
       will be 150 more than the original
       magic square. The total for each row,
       column and diagonal is 165.

   c i

−3	4	−1
2	0	−2
1	−4	3

   ii Each number is 5 less than the
       original number, therefore each row,
       column etc. that has three numbers
       will be 15 less than the original
       magic square. The total for each row,
       column and diagonal is 0.

6 Maximum safe mass is 94 kg

Exercise 1.2 (page 5)

1 a 750 b 980 c 640
   d 6500 e 360 f 1260
   g 1440 h 23 100 i 1440
   j 1280 k 18 900 l 21 600

2 a 31 b 21 c 400
   d 4 e 70 f 70
   g 10 h 60 i 60

3 400

4 7

5 281

6 18

Exercise 1.3 (page 7)

1 a

×	12	35	48	125
−2	−24	−70	−96	−250
−6	−72	−210	−288	−750
−12	−144	−420	−576	−1500

   b Other divisions are possible
      $-70 \div 35 = -2$
      $-96 \div 48 = -2$
      $-250 \div 125 = -2$ etc.

2     $12 \div 6 = 2$ $-18 \div -3 = 6$
      $12 \div 2 = 6$ $-18 \div -9 = 2$
      $-3 \times 6 = -18$ $2 \times -9 = -18$
      $-18 \div 6 = -3$ $-18 \div 2 = -9$

3 a 36 (rearranged)
   b −648 (36 changed sign)
   c 324 (18 is half of 36, therefore halve the
      answer)
   d −72 (9 is half of 18, therefore double 36
      also sign of 648 changed)
   e −162 (sign of one number changed. Also,
      both numbers halved therefore answer
      is quartered)
   f −36 (648 was doubled, therefore
      double 18)

Workbook answers

Exercise 1.1 (page 2)

1 a 80   b 60   c 90
   d 45   e 23   f 48
2 a 49 miles   b 98 miles
3 a 180   b 290   c 250
   d 70   e 210   f −40
4 a 3 glasses   b 120 ml
5 8 weeks
6 a 48   b 35

Exercises 1.2–1.3 (page 4)

1 a 1680   b 600   c 7200
   d 700   e 16 000   f 51 000

2 a 17   b 700   c 8
   d 3   e 120   f 120
3 144 loaves
4 258 legs
5 43 rows
6 a −65   b 330   c −9
   d −1000   e −800
7

×	−12	30	180
−7	84	−210	−1260
−25	300	−750	−4500

2 Properties of two-dimensional shapes

Prior knowledge

Students will already be able to recognise and name quadrilaterals. They will be able to classify quadrilaterals according to their properties including parallel sides, diagonals, symmetry and angle properties. Students will know that a circle is a 2D shape that has a centre, and will be able to identify and name the circumference, radius and diameter.

Objective overview

Learning objective	Objective code	*Student's Book* pages	*Workbook* pages	*Teacher's Guide* pages	Online resources
Identify, describe and sketch regular polygons, including reference to sides, angles and symmetrical properties.	7Gg.01	8–12	5–6; 53–56; 77–80	7–13	Flashcards Unit 2 Knowledge test Unit 2
Understand that if two 2D shapes are congruent, corresponding sides and angles are equal.	7Gg.02	8–12	5–6	7–13	
Know the parts of the circle: centre, radius, diameter, circumference, chord, tangent.	7Gg.03	8–12	5–6	7–13	

Background information

In this lesson, students will learn what a polygon is, and be able to identify types of common polygons by name and properties. They will examine polygon sides and lengths, vertices and angles. They will be able to identify rotational and reflective symmetry in shapes.

Much attention is paid to the parts of the circle in this lesson. Students will be able to locate and define each given part.

In this unit, students look at various 2D shapes and identify their properties, commenting on similarities and differences.

Terminology

There is much terminology for the students to learn in this lesson.

Students will be able to define a polygon and, in particular, a regular polygon and know the names of regular polygons such as pentagon, hexagon and octagon.

Students should learn that a four-sided polygon is known as a quadrilateral, and that there are many types of quadrilateral such as:

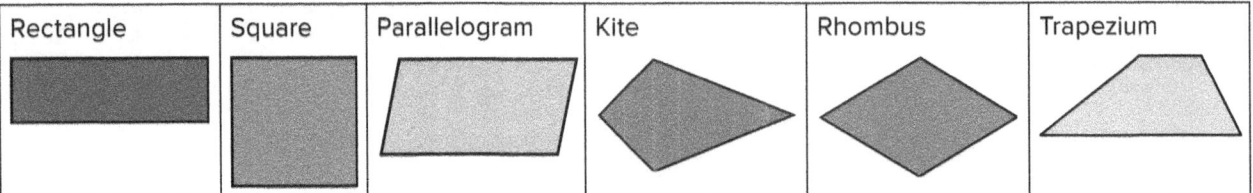

Rectangle	Square	Parallelogram	Kite	Rhombus	Trapezium

Students will also learn to name and identify parts of the circle. These parts are given in the *Student's Book* and repeated here for your reference.

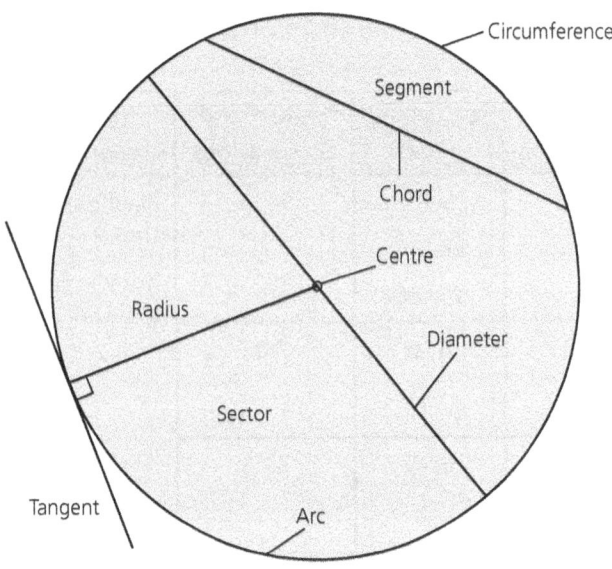

Lesson ideas

As an introduction, once the word polygon has been defined, students might be asked to draw polygons of sides 3–10. They will soon realise that it is difficult to draw regular polygons, especially 5-, 7- and 9-sided polygons.

The *Student's Book* asks students to complete a table of quadrilaterals. Teachers might want to pin up a large one for display. If this can be made by a student or small group so much the better. Similarly, one for regular polygons and symmetry is a good idea. The same goes for a large poster showing parts of a circle. The parts of the circle can be tested in pairs by students as a start of a lesson.

Starter activity

Scrambled words

Unscramble the words and decode the message.

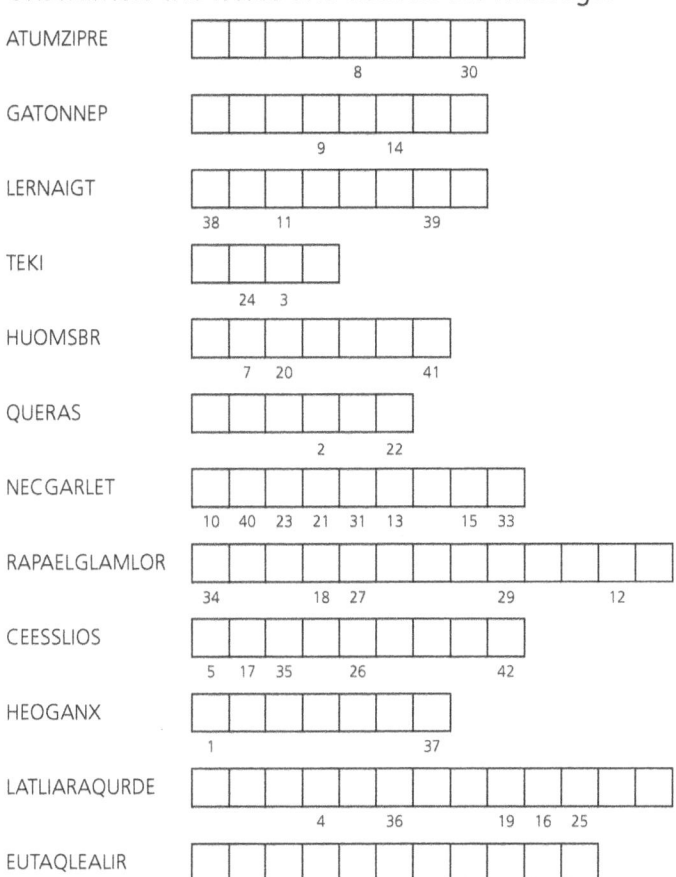

ATUMZIPRE

GATONNEP

LERNAIGT

TEKI

HUOMSBR

QUERAS

NECGARLET

RAPAELGLAMLOR

CEESSLIOS

HEOGANX

LATLIARAQURDE

EUTAQLEALIR

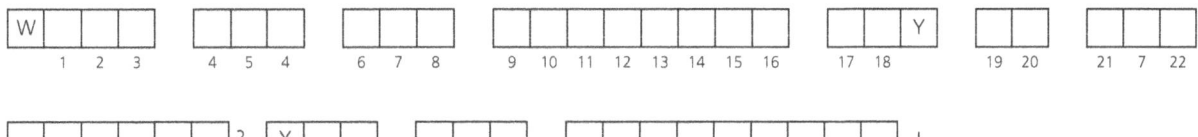

Answers

TRAPEZIUM	RECTANGLE
PENTAGON	PARALLELOGRAM
TRIANGLE	ISOSCELES
KITE	HEXAGON
RHOMBUS	QUADRILATERAL
SQUARE	EQUILATERAL

Message: What did the triangle say to the circle? You are pointless!

Activity

Shape shifter

This is a reinforcement activity which fits well after revision of polygon properties completed in Exercise 2.2.

You can join dots on a 3 by 3 square to make different quadrilaterals like this:

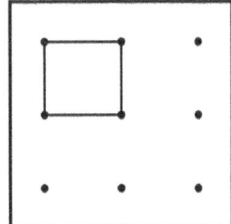 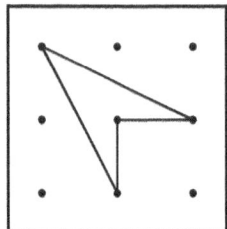

1   Use square dotty paper or a 9-pin geoboard and an elastic band to find as many DIFFERENT quadrilaterals as you can.
    What are the names of the quadrilaterals?
2   Choose a quadrilateral that you have found.
    Describe your quadrilateral to a friend.
    Can they draw your quadrilateral without looking at your shape?
    Here are some phrases you might find helpful:
    – equal sides   – line of symmetry   – right angles.
    – parallel sides   – rotational symmetry

Answers

There are 16 possible quadrilaterals.

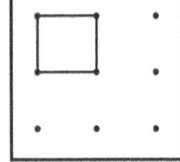 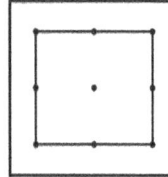 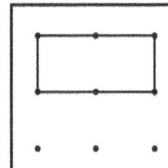 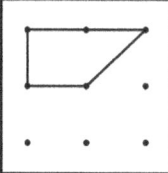

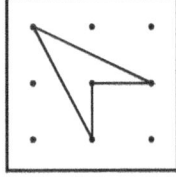 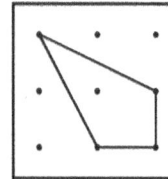 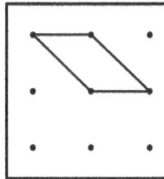 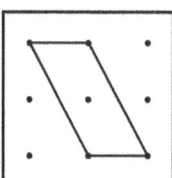

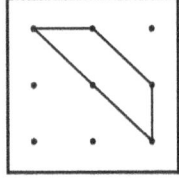 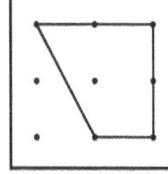 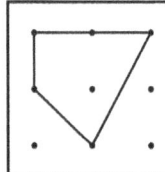 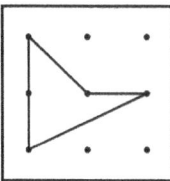

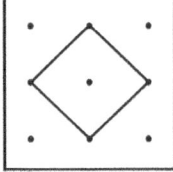  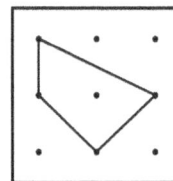 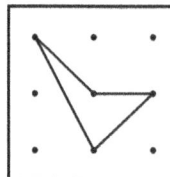

Student's Book answers

Exercise 2.1 (page 8)

1

	Rectangle	Square	Parallelogram	Kite	Rhombus	Trapezium
Opposite sides equal in length	Yes	Yes	Yes	No	Yes	No
All sides equal in length	No	Yes	No	No	Yes	No
All angles right angles	Yes	Yes	No	No	No	No
Both pairs of opposite sides parallel	Yes	Yes	Yes	No	Yes	No
Diagonals equal in length	Yes	Yes	No	No	No	No
Diagonals intersect at right angles	No	Yes	No	Yes	Yes	No
All angles equal	Yes	Yes	No	No	No	No

2 Square and parallelogram

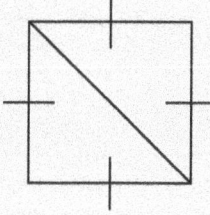 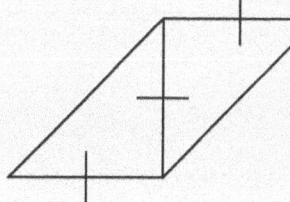

3 Rhombus

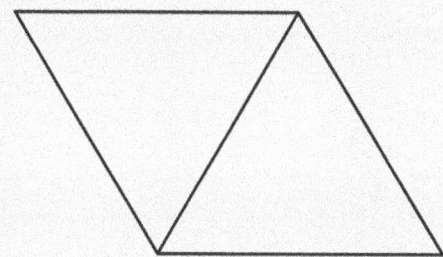

Exercise 2.2 (page 10)

1 a

Name of regular polygon	Number of sides	Shape	Number of lines of symmetry	Order of rotational symmetry
Equilateral triangle	3	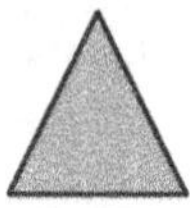	3	3
Square	4		4	4
Pentagon	5	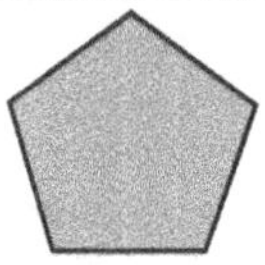	5	5
Hexagon	6	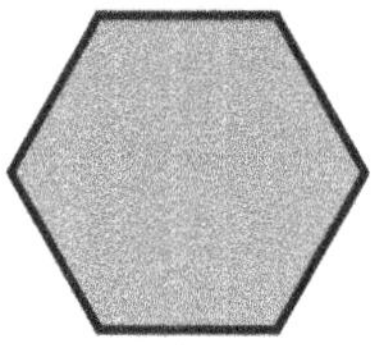	6	6
Octagon	8	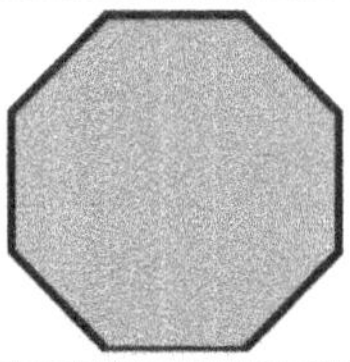	8	8
Decagon	10	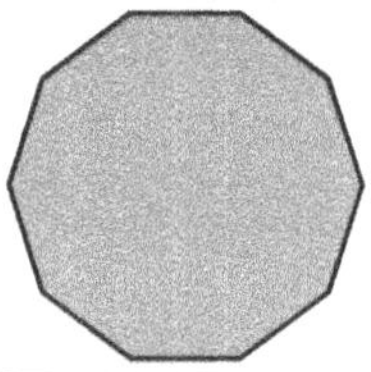	10	10

  b All the same.

2 a 1   **b** 1

3 a 3   **b** 3

4 a Octagon and square

  b Yes other combinations are possible. The smallest amount of sides a polygon can have in order to have rotational symmetry order 4 is 4 sides therefore a square is needed. The other polygon has to have a multiple of 4 for its number of sides, e.g. 8, 12, 16 etc.

  c

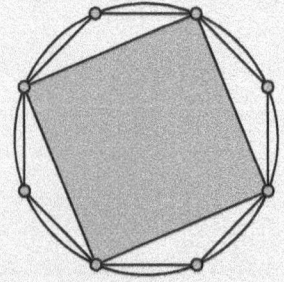

Exercise 2.3 (page 12)

1 a circle   **b** circumference   **c** radius / radii

  d chord   **e** diameter   **f** arc

  g sector   **h** segment   **i** tangent   **j** 90°

2 Other answers possible
  a The diameter is twice the length of the radius.
  b A diameter is a chord which passes through the centre of the circle.
  c An arc is the part of the circumference enclosed between two radii.
  d A sector and a segment are both parts of the area of a circle.

Workbook answers

Exercises 2.1–2.3 (page 5)

1 **a** Square

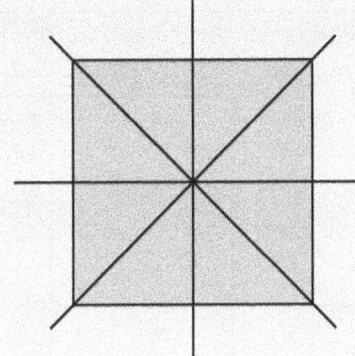

  b Rhombus

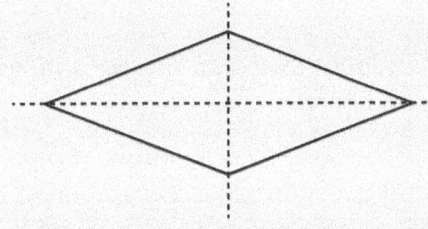

  c Trapezium or isosceles trapezium

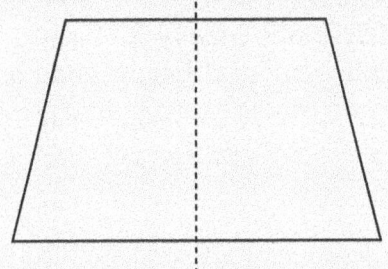

  d Hexagon or regular hexagon

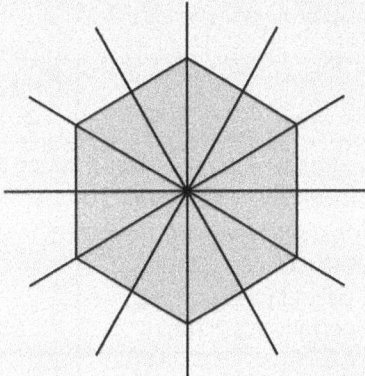

2 **a** chord   **b** segment
  c radius   **d** sector

3 **a** A parallelogram does not have any right angles.
  b A square has all equal sides or the diagonals in a square intersect at right angles.

4 **i, ii** Opposite sides are equal length and opposite sides are parallel.

5

6

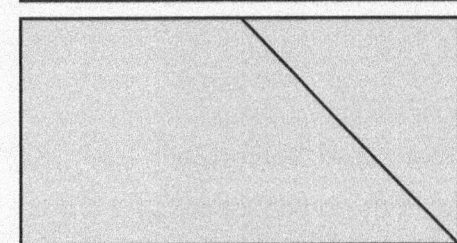

7 **a** Yes, a rectangle has 4 right angles and opposite sides must be equal length.
  b No, a square has to have all sides equal length.

3   Data collection and sampling

Prior knowledge

In prior learning, students will have had experience of collecting, recording and displaying data, and they will have planned and carried out investigations to respond to statistical questions.

Objectives overview

Learning objective	Objective code	*Student's Book* pages	*Workbook* pages	*Teacher's Guide* pages	Online resources
Select and trial data collection and sampling methods to investigate predictions for a set of related statistical questions, considering what data to collect (categorical, discrete and continuous data).	7Ss.01	13–18	7–8	14–17	Flashcards Unit 3 Knowledge test Unit 3
Understand the effect of sample size on data collection and analysis.	7Ss.02	13–18	7–8	14–17	

Background information

In this unit, students learn about different types of data. They learn that quantitative data (data which can be measured) falls into two categories – discrete and continuous.

Students will consider methods of data collection and how to ensure a representative sample. Data collection methods explored in this unit include interviews and questionnaires. Evaluation of the advantages and disadvantages of each data collection method is considered.

 For Unit 3 students will look at the different forms of data collection and reflect on their strengths and weaknesses in different situations. This can be linked to Cambridge Lower Secondary Science Stage 7 topic Thinking and Working Scientifically – Carrying out Scientific Enquiry. It also links to Cambridge Lower Secondary Global Perspectives challenges 'Globalisation – Global brands' and 'Employment – Why work?', where students are asked to create and conduct surveys and present the results.

Terminology

Discrete data – data that may only take specific values.

Continuous data – data that can take *any value* (usually within a range). The timing of an event is an example of continuous data.

Categorical data – data that falls into groups (or categories).

Lesson ideas

It is useful to begin with asking students if they believe information shown on graphs or charts more than in words. If so, why is this? Then ask if they can always trust these graphs and charts. The important idea is that the charts are accurate in illustrating the data collected but often that information is not given to us.

The *Student's Book* explains types of data well. This is a revision of Level 6 work. Teachers should emphasise 'sampling' and the *Student's Book* exercises make this clear.

Interviews and questionnaires: it is worth students working in groups on an interview. They should soon realise its limitations when they compare with other groups. The table in the *Student's Book* could be copied as a poster, by students if possible.

Starter activity

Wordsearch

Here are ten key words you will meet in this unit.

CATEGORICAL	CONTINUOUS	DATA	DISCRETE	POPULATION
QUANTITATIVE	QUESTIONNAIRE	RANDOM	REPRESENTATIVE	SAMPLE

Find the words hidden in the grid.

Y	L	F	D	B	V	U	K	O	V	U	C	R	Q
I	Q	A	V	B	V	U	X	R	N	A	N	F	U
R	H	U	S	U	O	U	N	I	T	N	O	C	A
X	A	C	E	W	N	A	Q	E	A	E	I	A	N
X	E	N	F	S	N	H	G	J	T	V	T	F	T
Y	R	E	D	N	T	O	I	E	O	M	A	W	I
M	D	X	R	O	R	I	R	C	V	W	L	S	T
K	A	Y	X	I	M	C	O	V	M	R	U	B	A
P	T	Y	C	I	S	I	J	N	X	N	P	Q	T
D	A	A	Z	I	P	R	F	G	N	H	O	H	I
T	L	W	D	S	A	M	P	L	E	A	P	H	V
R	E	P	R	E	S	E	N	T	A	T	I	V	E
Y	W	M	R	L	R	H	X	C	Z	B	R	R	S
Q	R	M	M	W	P	Z	K	D	X	B	N	Y	E

Answers

Y	L	F	D	B	V	U	K	O	V	U	C	R	Q
I	Q	A	V	B	V	U	X	R	N	A	N	F	U
R	H	U	S	U	O	U	N	I	T	N	O	C	A
X	A	C	E	W	N	A	Q	E	A	E	I	A	N
X	E	N	F	S	N	H	G	J	T	V	T	F	T
Y	R	E	D	N	T	O	I	E	O	M	A	W	I
M	D	X	R	O	R	I	R	C	V	W	L	S	T
K	A	Y	X	I	M	C	O	V	M	R	U	B	A
P	T	Y	C	I	S	I	J	N	X	N	P	Q	T
D	A	A	Z	I	P	R	F	G	N	H	O	H	I
T	L	W	D	S	A	M	P	L	E	A	P	H	V
R	E	P	R	E	S	E	N	T	A	T	I	V	E
Y	W	M	R	L	R	H	X	C	Z	B	R	R	S
Q	R	M	M	W	P	Z	K	D	X	B	N	Y	E

Student's Book answers

Exercise 3.1 (page 14)

1 a continuous
  b categorical
  c discrete
  d discrete
  e continuous
  f categorical
  g categorical
  h continuous
  i continuous
  j continuous
  k discrete
  l continuous

Exercise 3.2 (page 15)

Answers dependent on student's class.

Exercise 3.3 (page 15)

1 a They will all be the same age (also might be the same gender/similar ability). Depending on the class size, it might not be very many to represent 1500 students.

  b Random sampling across all year groups.

2 a As they are just from her classes, they will only be responding based on her teaching not the teaching across the school.

  b By randomly selecting a few students from each maths class in the school.

3 a There are far more females under 14 than there are people in any other group so they will be underrepresented in the sample. In contrast, as there are only 5 males over 14, their views will all be represented in the survey.

  b As there were 20 people in the sample, and there are 100 people in the population, they should randomly select $\frac{1}{5}$ of each group. The number of people sampled from each group is therefore:

	Male	Female
Under 14	4	12
Over 14	1	3

Exercise 3.4 (page 16)

1 a The student's answers are very variable and are dependent on other factors such as friends, what lessons he has had in the morning etc.
   b Too many questions asked at the same time.
   c Students' answers will vary, e.g. On average how often do you eat school food a week?
   d Students' answers will vary, e.g. Answers are very detailed, interviewer can ask more questions if answers need clarifying.
   e Time consuming.

2 a Students' answers will vary, e.g. Easy to analyse. Easy to hand out to a lot of people.
   b Students' answers will vary, e.g. No opportunity to give detailed answers. Students may not hand the questionnaires back in.
   c Students' answers will vary.
   d 20 is not a big sample, so the responses may not be representative of the whole school.
   e Students' answers will vary. There is no correct answer to this as the representativeness of the sample is more important than the sample size in itself.

Workbook answers

Exercises 3.1–3.2 (page 7)

1 a continuous   b categorical
   c discrete   d continuous
2 a discrete   b continuous
   c categorical   d population
   e sample
   f Quantitative data is data that can be measured; quantitative data can either be discrete or continuous.
3 Students' answers will vary.

Exercise 3.3 (page 8)

1 a No time frame given, e.g. week? month? year? etc.
   b i, ii No option for 'never', no option '4'.
   c Standing outside the gym is not likely to give a sample representative of the whole population as all the respondents will be gym goers.
2 Students' answers will vary, e.g. How many portions of fruit and/or vegetables do you eat per day?
   • None
   • 1
   • 2
   • 3
   • 4
   • 5 or more

Area of a triangle

Prior knowledge

Students will already be able to recognise and name rectangles and triangles. Students will know how to find the area of rectangles and will be able to use this to work out the area of right-angled triangles.

Objectives overview

Learning objective	Objective code	Student's Book pages	Workbook pages	Teacher's Guide pages	Online resources
Derive and know the formula for the area of a triangle. Use the formula to calculate areas of triangles and compound shapes made from rectangles and triangles.	7Gg.05	19–24	9–11	18–21	Flashcards Unit 4 Knowledge test Unit 4 Worksheet: Fenced off

Background information

In this unit, students will learn how to calculate the area of a triangle. They will use the learned formula and apply it to different types of triangle. Students will also consider how to use their knowledge to calculate the area of compound shapes.

Terminology

When calculating the area of triangles which do not contain a right angle, it is important to clarify that the **perpendicular height** (that is, the height at right angles to the side chosen as the base) is used to multiply by the base, before being halved.

Lesson ideas

Draw a few rectangles on the board and put on dimensions. Ask students to give the area. Then ask other students to draw the diagonal and say what the area of the right-angled triangle is. This should be enough revision. Some mental questions might be set too if necessary.

The *Student's Book* is very clear on how to derive the formula for the area of a triangle. The exercises are clear and students should enjoy looking at the Sierpinski triangles. Finding the area of compound shapes is clearly explained and there are plenty of questions for students to answer in the *Student's Book* and *Workbook*.

Starter activity

Triangle Tangle

You can join dots on a 3 by 3 square to make different triangles like this:

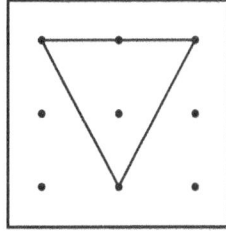 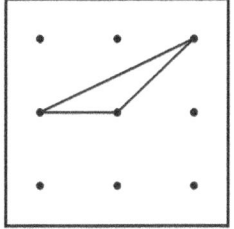

1. Use square dotty paper or a 9-pin geoboard and an elastic band to find as many DIFFERENT triangles as you can.
2. By counting squares, work out the area of each triangle.
   One small square has an area of 1 cm².

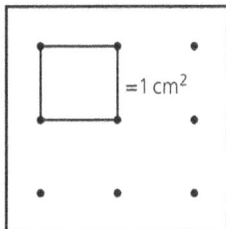

Answers

1 There are eight possible triangles.

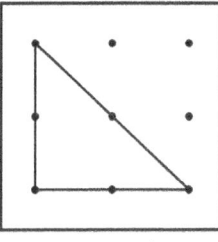 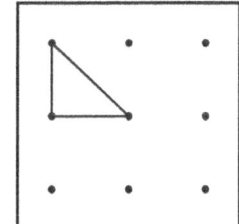 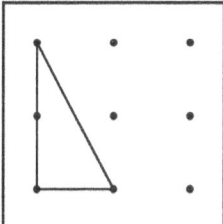

Area = 2 cm²   Area = 0.5 cm²   Area = 1 cm²

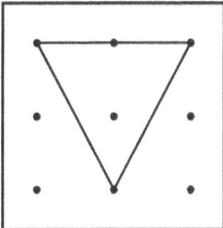 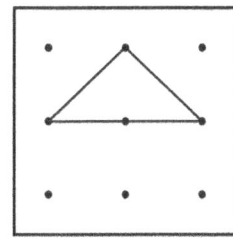 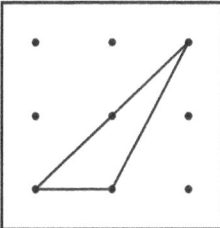

Area = 2 cm²   Area = 1 cm²   Area = 1 cm²

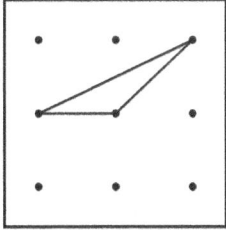 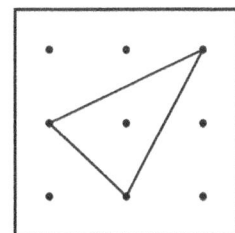

Area = 0.5 cm²   Area = 1.5 cm²

TWM activity notes

Exercise 4.1 is explained here as an exemplar of Thinking and Working Mathematically (TWM), detailed in the Introduction to the *Teacher's Guide*, page xi. Students have just covered how to calculate the area of a triangle using the formula and have been told of the need for the height to be measured at right angles to the base.

Q3   *Three triangles P, Q and R share a common base and lie between two parallel lines.*

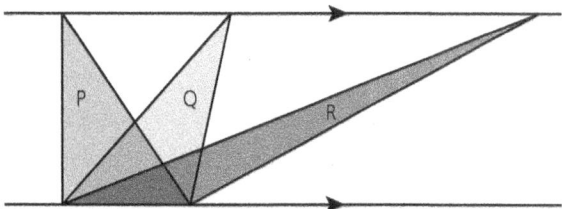

         Which of the three triangles (if any) has the biggest area? Give a reason for your answer.

Here the actual value for the area is not needed as no measurements are given. Students need to identify the mathematical properties common to each of the triangles (they share the same base and also have the same perpendicular height) and therefore generalise that triangles with the same base length and perpendicular height will have the same area, regardless of shape.

TWM characteristics: Convincing   Characterising

This question if written in a standard way could have been presented as follows:

Q3   *Three triangles P, Q and R have a base length of 6 cm and a perpendicular height of 10 cm as shown:*

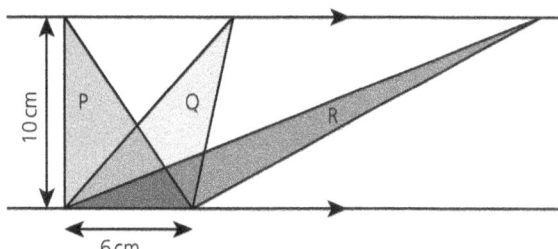

        Calculate the area of each triangle. Comment on your answers.

Students here are required to apply the formula to each of the three triangles and then conclude that they are all the same.

Student's Book answers

Exercise 4.1 (page 21)

1 a 6 cm²   b 20 cm²
   c 55 cm²
2 a 17.28 cm²   b 10 cm
   c 2 cm   d 10 cm
3 All triangles have the same area as the perpendicular height is equal for each triangle and they each share the same base.
4 a Answers will vary depending on drawing. Height is shown by the dashed line.

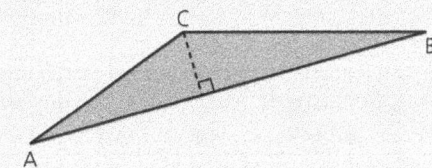

   b Answers will vary depending on drawing. Height is shown by the dashed line.

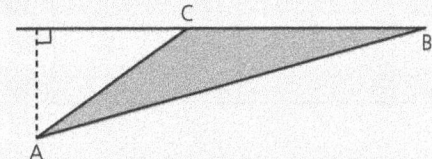

   c Answers will vary depending on drawing. Height is shown by the dashed line.

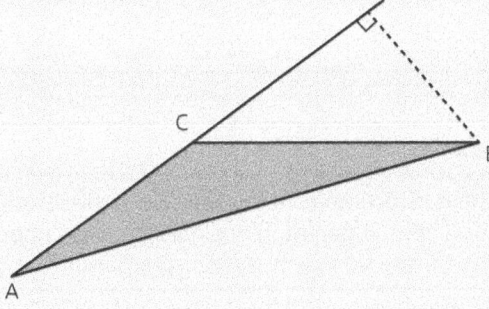

   d Equal. Area of each triangle should be equal unless they were drawn/measured inaccurately.

5 a R can be at any point along the dark line shown.
   b R can be at any point along the dark line shown.

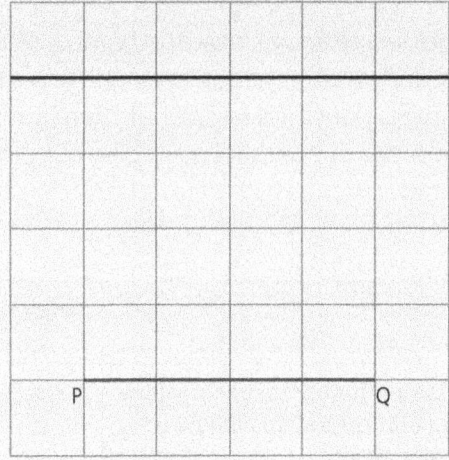

   c All are 4 units above the base PQ.
6 2: 48 cm²   3: 36 cm²   4: 27 cm²
   5: 20.25 cm²

Exercise 4.2 (page 23)

1 90 cm²
2 104 cm²
3 300 cm²
4 $x = 4$ cm
5 a $\dfrac{5}{8}$ or equivalent.
   b Each triangle needs to have an area of 32 cm² and the sum of their perpendicular sides equal to 16 cm, e.g. area $= \dfrac{1}{2} \times 8 \times 8$
   c No as the triangles would start to overlap.

Workbook answers

Exercise 4.1 (page 9)

1 a 24 m²   b 22 mm²   c 7.5 cm²
2 a 9.69 cm²   b 8 mm   c 12 m
   d 7 cm
3 a 3.75 cm   b 2 cm

Exercise 4.2 (page 11)

1 a 78 cm²   b 165 cm²
2 $560

Order of operations

Prior knowledge

Students will already know and be able to apply laws of arithmetic to calculations. They will understand what it means to simplify a calculation and will be able to use laws of arithmetic to do so. Students will know that there is an order of mathematical operations and will be able to use the order of operations to solve calculations.

Objectives overview

Learning objective	Objective code	*Student's Book* pages	*Workbook* pages	*Teacher's Guide* pages	Online resources
Understand that brackets, positive indices and operations follow a particular order.	7Ni.02	25–27	12–13	22–25	Flashcards Unit 5 Knowledge test Unit 5 Worksheet: Four fours

Background information

In this unit, students learn that the order in which mathematical calculations are done depends on the operations being used.

The given order is:

Brackets	Any operation in brackets is done first.
Indices	A number raised to a power (index) is done next.
Division and/or **M**ultiplication	Multiplications and divisions are done next. Starting from the left, work these out in the order that they appear in the equation. If multiplication appears first you should complete this before division.
Addition and/or **S**ubtraction	Additions and subtractions are carried out last. Start from the left and work these out in the order that they appear in the equation. If subtraction appears before addition, you should complete this first.

A way of remembering this order is with the shorthand **BIDMAS**.

Terminology

The index of a number says how many times to use the number in a multiplication. It is written as a small number to the right and above the base number.

In this example: $8^2 = 8 \times 8 = 64$

The plural of index is indices. (Other names for index are exponent or power.)

Lesson ideas

It is useful to have a poster showing BIDMAS which a student might make.

After the worked example in the *Student's Book* has been looked at simple questions like: $1+4\times5$ could be set as mental questions before students work through the exercises.

The four 4's question is quite difficult, but the brightest students will enjoy the challenge. Tell them not to give up. Not solving a puzzle is often as valuable as solving one. Perhaps mention Fermat's Last Theorem: Andrew Wiles worked on it for 12 years on and off before he solved the problem!

Starter activity

 Target 100

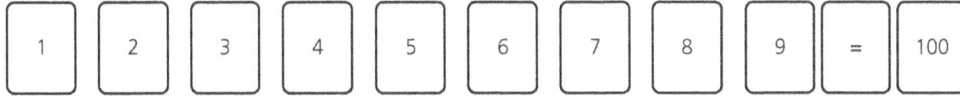

1 Use only the operations + and × between the number cards to make the target of 100.
2 Can you find other ways to reach 100?
   You can use +, −, ÷ and × and brackets if you need to.
   You can put the cards together to make a 2-digit number, but the order must be the same.

Answers

1 $1+2+3+4+5+6+7+8\times9=100$
2 Many possible answers.

Activity

 One, Two, Three, Four...

This activity is similar to the four 4's question. However, it can be answered without the need for squaring or use of other functions. Less confident students can be challenged to find the numbers from 1 to 10 – only 7 requires the use of brackets.

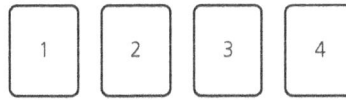

As a challenge activity, students can use the cards to find the numbers from 21 to 50.

Make all the numbers from 1 to 20 using all four cards, brackets and the operations +, −, ÷ and ×.

For example:

$$2 \times 3 - 4 + 1 = 3$$

Write out your calculations using brackets.

Answers

$2 \times 3 - 4 - 1 = 1$ $2 \times 3 + 4 + 1 = 11$

$4 - 3 + 2 - 1 = 2$ $3 \times 4 \times (2 - 1) = 12$

$2 \times 3 - 4 + 1 = 3$ $3 \times 4 - 1 + 2 = 13$

$2 \times 4 - 3 - 1 = 4$ $1 \times 3 \times 4 + 2 = 14$

$2 \times 4 - 1 \times 3 = 5$ $3 \times 4 + 1 + 2 = 15$

$2 \times 4 - 3 + 1 = 6$ $2 \times (1 + 3 + 4) = 16$

$3 \times (4 - 1) - 2 = 7$ $3 \times (2 + 4) - 1 = 17$

$2 + 3 + 4 - 1 = 8$ $1 \times 3 \times (2 + 4) = 18$

$2 \times 3 + 4 - 1 = 9$ $3 \times (2 + 4) + 1 = 19$

$1 + 2 + 3 + 4 = 10$ $1 \times 4 \times (2 + 3) = 20$

Student's Book answers

Exercise 5.1 (page 26)

1 a 9   b 24   c 6   d 0   e −18
   f 39   g 11   h 9   i 16   j 16

2 a Carla is right as the squaring is done before the multiplication.
   b Ibrahim carried out the multiplication first and then squared the answer, which is wrong.

3 a $(3 + 2) \times 5 + 4 = 29$ b $3 + 2 \times (5 + 4) = 21$

4 a $8 + 6 - 4 = 10$ b $8 - 6 + 4 = 6$
   c $7 + 6 \div 3 = 9$ d $7 \times 6 - 3 = 39$
   e $7 \times (6 - 3) = 21$ f $8 \div (4 - 2) = 4$
   g $(8 - 4) \div 2 = 2$ h $2 \times (3 + 5 + 2) = 20$
   i $2 \times (3 + 5) + 2 = 18$ j $(3 + 9) \times (8 - 5) = 36$

5 a Brackets not necessary as the multiplication would still be carried out before the addition.
   b Brackets needed as the addition needs to be carried out before squaring in order for the calculation to be correct.
   c Brackets not needed as the squaring would still be carried out first without the brackets and the −6 and +8 can then be carried out in either order.

6 a i $(4 + 4) \div (4 + 4) = 1$ ii $4 \times 4 \div 4 \div 4 = 1$
         other solutions are possible   other solutions are possible
   b $4 \div 4 + 4 \div 4 = 2$
      other solutions are possible
   c i The solutions presented here are just one example of each. Other solutions are possible.

$(4 + 4 + 4) \div 4 = 3$	$4 \div 4 + 4 + 4 = 9$	$4 \times 4 - 4 \div 4 = 15$
$(4 - 4) \times 4 + 4 = 4$	$4 \times 4 - 4 - \sqrt{4} = 10$	$4 + 4 + 4 + 4 = 16$
$(4 \times 4 + 4) \div 4 = 5$	$44 \div \sqrt{4} \div \sqrt{4} = 11$	$4 \times 4 + 4 \div 4 = 17$
$(4 + 4) \div 4 + 4 = 6$	$4 \times (4 - 4 \div 4) = 12$	$4 \times 4 - \sqrt{4} + 4 = 18$
$4 + 4 - 4 \div 4 = 7$	$44 \div 4 + \sqrt{4} = 13$	$4 - 4 \div 4 + 4^2 = 19$
$4 + 4 + 4 - 4 = 8$	$4 \times 4 + \sqrt{4} - 4 = 14$	$(4 + 4 \div 4) \times 4 = 20$

   ii Students' solutions

Workbook answers

Exercise 5.1 (page 12)

1 a 6 b 29 c 22
  d 7 e 22 f 53
  g 6 h 7 i 24
  j 27

2 a Joe is correct. $2^2 = 4$, $5 \times 4 = 20$
  b Thea did 5×2 first and then squared her answer.

3 a 19 b 24 c 35
  d 9

4 Her method is incorrect. Firstly, $3^2 = 9$, she multiplied the 3 by 2 rather than squaring. Secondly, she added $6 + 4$ first, she should have multiplied 4 and 2. Her answer should have been 17.

5 a $7 + 3 \times (5 - 2) = 16$
  b $(3 + 5)^2 \div 2 = 32$
  c $4 \times (3 + 2 \times 5) = 52$
  d $18 \div (2 + 4) \times 5 = 15$

6 a $8 + 7 - 3 + 1 = 13$
  b $2 \times 5 + 3 \times 4 = 22$
  c $50 \div 5 - 2 \times 3 = 4$
  d $(6 + 3) \times 4 - 10 = 26$

7 a $3 \times 2 + 7 = 13$
  b $4 + 2 \times 5 = 14$
  c $6 \div 2 + 3 = 6$
  d $10 - 6 - 3 = 1$
  e $4 \times (2 + 5) = 28$
  f $(10 - 5) \times (3 + 1) = 20$
  g $3 \times (4 - 7 \times 2) = -30$

Algebra beginnings – Using letters for unknown numbers

6

Prior knowledge

Students will need to be confident that they understand the order of operations as it has been taught in Unit 5.

Objectives overview

Learning objective	Objective code	*Student's Book* pages	*Workbook* pages	*Teacher's Guide* pages	Online resources
Understand that letters can be used to represent unknown numbers, variables or constants.	7Ae.01	28–35	14–17	26–29	Flashcards Unit 6
Understand that the laws of arithmetic and order of operations apply to algebraic terms and expressions (four operations).	7Ae.02	28–35	14–17	26–29	Knowledge test Unit 6

Background information

In this unit, students learn that in algebra letters are used to represent unknown numbers. Often the task is to find the value of the unknown number, but not always.

Students will learn to define, construct and simplify algebraic expressions (using BIDMAS learned in the previous unit). They will substitute given values into expressions.

Students will derive formulae and calculate answers using them.

 You could draw parallels between using formulae in Science, in line with Cambridge Lower Secondary Science and the use of symbols and formulae to represent scientific ideas.

Terminology

It is important that students are clear about the meaning of the terms expression, equation and formula. Students can work in pairs to give examples of each.

Lesson ideas

Students will be interested to know that algebra came from the book *Al Jabr* by Muhammad ibn Musa Al Khwarizmi in Baghdad around 830CE. This is a good way to introduce algebra. It helps students to appreciate both the history and international nature of mathematics. It is important that students in all countries are aware of the major contribution made by Arab scholars to all learning, but in particular, to mathematics and science. It translates as 'a book on calculation by completion and balancing.'

Order of operations in algebra may require a little revision of order of operations in arithmetic, but, since it is the preceding unit, not much should be needed.

The *Student's Book* and *Workbook* give plenty of questions involving substitution of numbers for letters in expressions and formulae.

Starter activity

Mind reading

Try this mind-reading trick:

- Think of a number
- Double it
- Add 12
- Divide by 2
- Subtract the number you first thought of
- The answer is 6

How does this trick work? Can you design your own puzzle?

Answer

The reason the answer is always 6 is clear when you use algebra.

• Think of a number	x
• Double it	$2x$
• Add 12	$2x + 12$
• Divide by 2	$x + 6$
• Subtract the number you first thought of	6

Support activity

Matching pairs

This activity would support students who need some additional reinforcement after working through Exercises 6.1 and 6.2.

Match together algebra cards which contain equivalent expressions. Which expressions don't have a matching pair?

Find the value of the expression on each card when a) $n = 3$ and b) $n = 4$.

$3 + n$	$2n + 6$	$4n - n$
$2n + n$	$2n + 3$	$n + n + n$
$6n$	$n - 3$	$3 - n$
$n + 2 + 2n + 1$	$2 + n + 1$	$3n$
$3 \times 2n$	$6n \div 2$	$n + 1 + 2$
$6 \times n$	$3 + n$	$n + 3$
$6n - 3n$	$n + 3 + n + 3$	$3 \times 2 \times n$
$n - 1 - 2$	$n + 3 - 2n$	$n + 2n + 3n$

Answers

$3 \times 2n = 6 \times n = 3 \times 2 \times n = n + 2n + 3n = 6n$;	a 18	b 24
$3n = 6n \div 2 = 6n - 3n = 3 \times n = 4n - n = 2n + n = n + n + n$;	a 9	b 12
$3 + n = n + 1 + 2 = n + 3 = 2 + n + 1$;	a 6	b 7
$n + 3 + n + 3 = 2n + 6$;	a 12	b 14
$n + 3 - 2n = 3 - n$;	a 0	b −1
$n - 1 - 2 = n - 3$;	a 0	b 1
$2n + 3$ doesn't have a matching pair;	a 9	b 11
$n + 2 + 2n + 1$ doesn't have a matching pair;	a 12	b 15

Student's Book answers

Exercise 6.1 (page 29)

1 a $7 + x$ b $2x + 1$
   c $2x + y$ d $4x + 2y + 3$
2 a i $x + x + 4 + 4$
     ii $2x + 8$
   b i $x + x + y + y$
     ii $2x + 2y$
   c i $2 + m + m + 2 + m + m + 2 + 2 + m + m$
     ii $8 + 6m$
   d i $y + y + y$
     ii $3y$
   e i $q + q + q + q + p + p + p + p$
     ii $4q + 4p$
   f i $y + 3 + y + 2 + y + 3 + y + 2$
     ii $4y + 10$
   g i $m + 8 + m + 2 + m + 8 + m + 2$
     ii $4m + 20$
   h i $y - 1 + x + x + 5 + 3 + 5 + y - 4$
     ii $2y + 2x + 8$
3 a Both blocks next to each other.
   b One blue block and two orange blocks next to each other.
   c Blue block with an orange block on top of it, the length of blue that is left uncovered.
   d Two blue blocks and two orange blocks next to each other.
4 a $x + y$ b $y - x$
   c $y - 10$ d $2x$
   e $x + y + 10$ f $y - x$
   g $y + x$
5 a Area $= 24x$, perimeter $= 12x + 8$
   b Partly correct, the area will be half but the perimeter will not be.
6 a $24a$ b $32a$

   c Students' explanations, e.g. there are more of the edges exposed in the 1st diagram compared to the 2nd.
   d $8a$

Exercise 6.2 (page 32)

1 a a b $3b$
   c $5c - 2b$ d $3d + e - 3f$
2 a $11j$ b $7h$
3 a Students' explanations
   b Students' explanations
   c Yes as it simplifies to $10x - 6$

Exercise 6.3 (page 33)

1 a 12   b −1
   c −6   d 5
2 a 8   b 2
   c 18   d 1

Exercise 6.4 (page 34)

1 a Perimeter $= 22$ cm   Area $= 28$ cm²
   b Perimeter $= 40$ cm   Area $= 96$ cm²
   c Perimeter $= 13$ cm   Area $= 9$ cm²
   d Perimeter $= 20.5$ cm   Area $= 18$ cm²
   e Perimeter $= 81.6$ cm   Area $= 32$ cm²
   f Perimeter $= 3.4$ cm   Area $= 0.6$ cm²
   g Perimeter $= 2.9$ m   Area $= 0.45$ m²
   h Perimeter $= 12.6$ m   Area $= 2.9$ m²
2 a Students' explanations
   b Neither is correct. If $x > 1.5$ then
     $6x + 4 > 2x + 10$
     If $x < 1.5$ then $2x + 10 > 6x + 4$
3 a 420 V   b 3200 V
   c 600 V   d 400 V

Workbook answers

Exercises 6.1–6.2 (page 14)

1 a $2x+10$ b $3a+8b$
   c $6y+3$ d $8f-2g$

2 a $6a+8$ b $9x+5$
   c $9y-3$ d $20b-3$

3 a

$24a+16$

$12a+6$	$12a+10$

$5a+2$	$7a+4$	$5a+6$

$2a+3$	$3a-1$	$4a+5$	$a+1$

   b

$16a+29$

$10a+18$	$6a+11$

$7a+6$	$3a+12$	$3a-1$

$3a-2$	$4a+8$	$4-a$	$4a-5$

4 a $5b$ b $3a+6$
5 a $4a^2+15a+3$ b $6b^2-b$

Exercise 6.3 (page 17)

1 a 9   b 24
   c 19   d 30
   e 22   f −15

2 a 7   b 1
   c −2   d 36
   e 3   f 40

3 a i 60 miles   b i 40 mph
      ii 32.5 miles   ii 70 mph
      iii 30 miles   iii 5 m/s

7 Organising and presenting data

Prior knowledge

In prior learning, students will have had experience of collecting, recording and displaying data. They will be able to understand and interpret a variety of statistical tables and diagrams including Venn diagrams, Carroll diagrams, tally charts, waffle diagrams, frequency tables, bar charts and pie charts. Students will have had experience constructing statistical diagrams including scatter graphs, dot plots and line graphs. Students will be able to select the most appropriate diagram to use for different situations.

Objectives overview

Learning objective	Objective code	*Student's Book* pages	*Workbook* pages	*Teacher's Guide* pages	Online resources
Record, organise and represent categorical, discrete and continuous data. Choose and explain which representation to use in a given situation: Venn and Carroll diagrams; tally charts, frequency tables and two-way tables; dual and compound bar charts; waffle diagrams and pie charts; frequency diagrams for continuous data; line graphs; scatter graphs; infographics.	7Ss.03	36–53	18–21; 92	30–36	Flashcards Unit 7 Knowledge test Unit 7

Background information

This unit builds upon the themes explored around data collection in Unit 3. Students build on this learning by constructing tally and frequency tables as simple ways of recording results. They will consider the use of two-way tables when dealing with more than one variable of results.

Students will then consider ways of displaying data visually, using such methods as:

- Venn diagrams
- Carroll diagrams
- bar graphs
- frequency diagrams
- compound bar charts

- pie charts
- waffle diagrams
- line graphs
- scatter graphs
- infographics.

 In this unit, students are introduced to the most common ways of collecting and representing data through tables, graphs and charts. This links to Cambridge Lower Secondary Science Stage 7 Thinking and Working Scientifically – Carrying out scientific enquiry and Scientific enquiry: analysis, evaluation and conclusions. It also links to Cambridge Lower Secondary Global Perspectives challenges 'Globalisation – Global Brands' and 'Employment – Why work?', where students are asked to create and conduct surveys and present the results.

Terminology

As demonstrated in the information given above, this unit contains many types of data collection and presentation. Students will need to be familiar with each one – when it is appropriate to use and how to construct it using the data collected. Each data collection and presentation method is defined in the glossary, provided within the online resources.

Lesson ideas

A quick revision of Unit 3 on data collection and sampling may be a good introduction to this unit. The worked example of a tally chart and of grouped data is clearly explained. A Venn diagram makes a good poster that a student might make. There is an opportunity to make plenty of display work from this unit either from the worked examples or from the exercises in the *Student's Book*. This is an opportunity for group work.

Starter activity

 ### Finding charts

Your next lesson is on organising and presenting data.

Before your next lesson look for examples of charts and diagrams used in newspapers, adverts and shops.

Bring any examples you can to the lesson.

Choose one of the charts or diagrams that you found and write down the answers to these questions:

- Is the chart eye-catching?
- Is the chart easy to read and understand?
- What is the author trying to show you with the chart?
- Why did they choose to use that particular chart?

Extension activity

 ### Statistics investigation

This stretch-and challenge activity is best used upon completion of all exercises in the *Student's Book* for Unit 7.

Plan your own statistics investigation.

Choose one of the following ideas or make up one of your own!

- Which country has the best athletes?
- Which is the best football team in the world?
- Do girls have faster reaction times than boys?
- Which is the most popular type of pet at your school?
- Which country is the wettest/driest/hottest/coldest/windiest?

Your investigation should have:

- a hypothesis – this is a statement that you wish to test; for example, 'Girls have a faster reaction time than boys'
- what your population is
- how you will take a sample and how large should your sample be
- what type of data you need to collect and how you will record it
- a chart or diagram to display your data
- a conclusion.

Student's Book answers

Exercise 7.1 (page 39)

1 a 19 days   **b** 3 days

2 a Discrete, because the accuracy of the measurement is controlled by the accuracy of the equipment, e.g. a length recorded as 12.3 cm could be 12.28367 etc. with more accurate tools.

c i 23
  ii 27

b

Length (cm)	Tally	Frequency
10–12	卌 ‖‖	8
12–14	卌 卌 卌	15
14–16	卌 卌 卌 ‖	17
16–18	卌 ‖	6
18–20	‖‖‖	4

3 a

	Lion	Camel	Monkey	Mouse	Other	Total
Boys	5	1	**10**	2	6	24
Girls	**2**	4	**5**	5	0	**16**
Total	7	**5**	**15**	**7**	**6**	40

b 2   **c** 10

d A boy. A rabbit falls in the 'other' category and only boys are in this category.

Exercise 7.2 (page 41)

1 a 42   **b** 22   **c** 5

2 a No, it is not correct because 12 members like both and so have been counted twice.

b 13

3 a Because there are 5-sided yellow shapes and 4-sided blue shapes and they cannot be placed in only one cell of the Carroll diagram.

b

	4-sided	5-sided
Blue	4, 5, 7, 10	3
Yellow	1, 9	2, 6, 8

Exercise 7.3 (page 45)

1

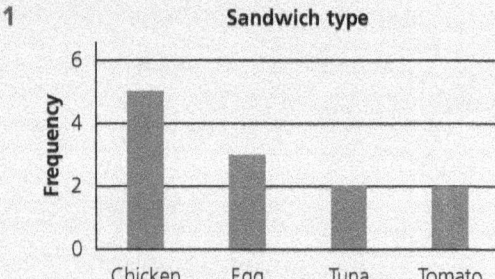

Sandwich type

2 a Students' answers will vary.

b, c Students' own data collection and frequency diagrams.

3 a 8

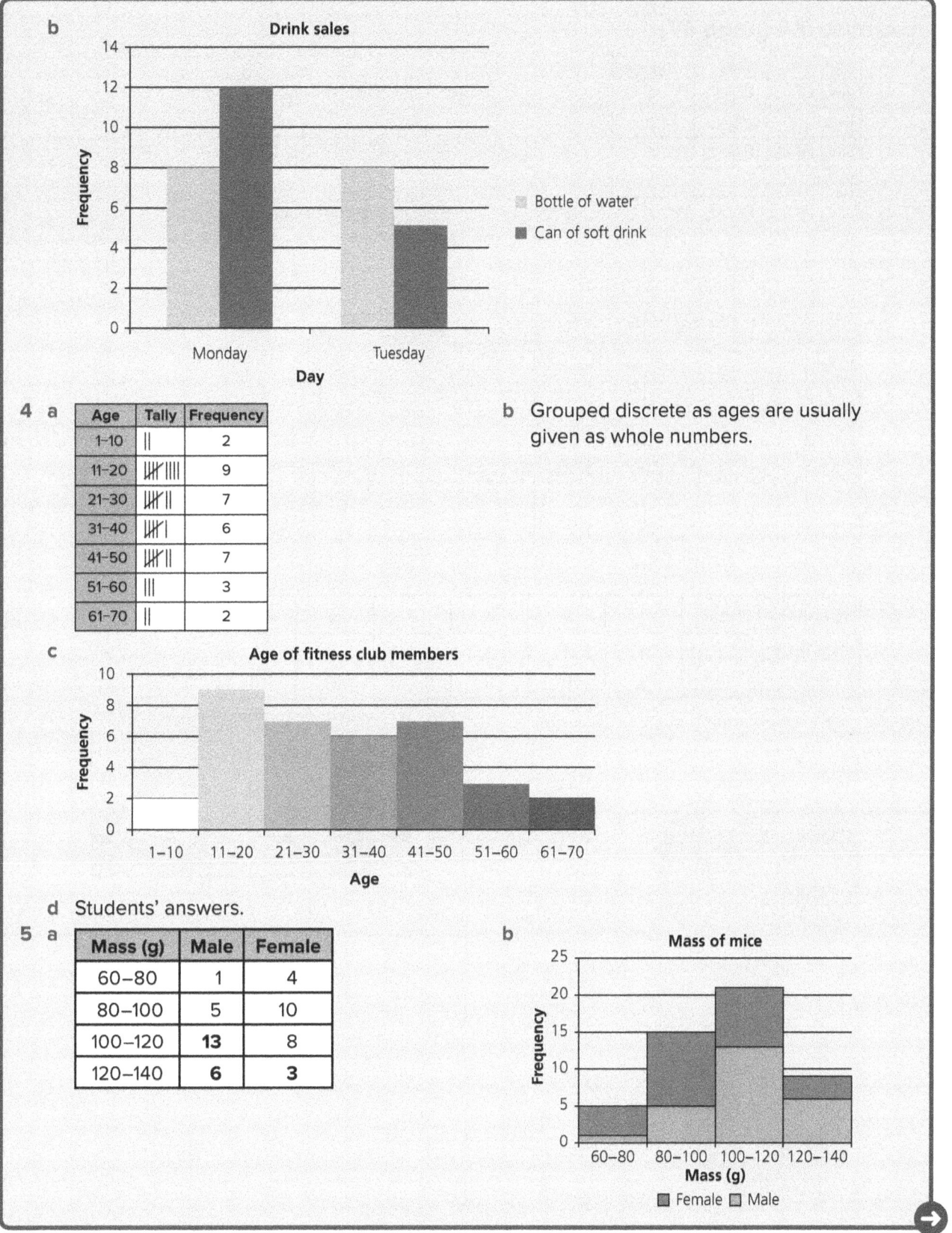

b

Drink sales

4 a

Age	Tally	Frequency
1–10	II	2
11–20	IIII IIII	9
21–30	IIII II	7
31–40	IIII I	6
41–50	IIII II	7
51–60	III	3
61–70	II	2

b Grouped discrete as ages are usually given as whole numbers.

c **Age of fitness club members**

d Students' answers.

5 a

Mass (g)	Male	Female
60–80	1	4
80–100	5	10
100–120	**13**	8
120–140	**6**	3

b **Mass of mice**

Exercise 7.4 (page 47)

1 a

	Tally	Frequency
Brown	IIII IIII II	12
White	IIII IIII	10
Wholemeal	IIII IIII IIII	15
Soda	III	3

b Brown $= \dfrac{12}{40} = \dfrac{3}{10}$

White $= \dfrac{10}{40} = \dfrac{1}{4}$

Wholemeal $= \dfrac{15}{40} = \dfrac{3}{8}$

Soda $= \dfrac{3}{40}$

c

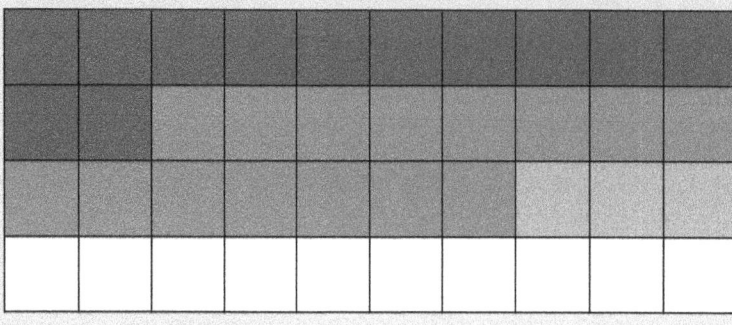

■ Brown   ■ Wholemeal   ■ Soda   □ White

2 a

Number of litres	1	2	3	4	5	6
Fraction	$\dfrac{1}{8}$	$\dfrac{1}{8}$	$\dfrac{3}{8}$	$\dfrac{2}{8} = \dfrac{1}{4}$	$\dfrac{1}{16}$	$\dfrac{1}{16}$

b Fraction of number of litres

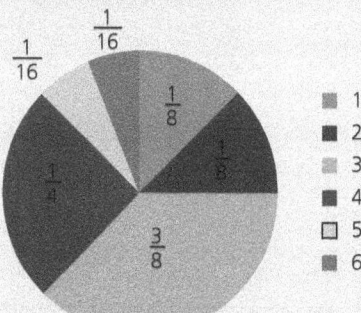

■ 1
■ 2
■ 3
■ 4
□ 5
■ 6

c

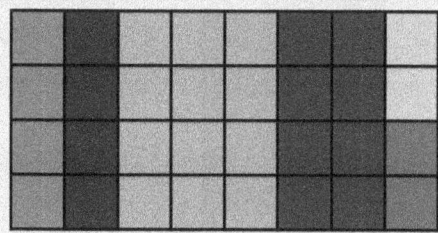

3 a Exam difficulty

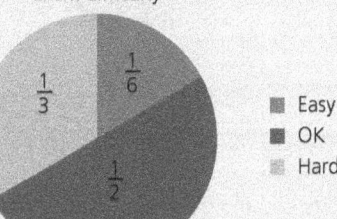

■ Easy
■ OK
■ Hard

b 15

4 a 1 – C
2 – D
3 – A

b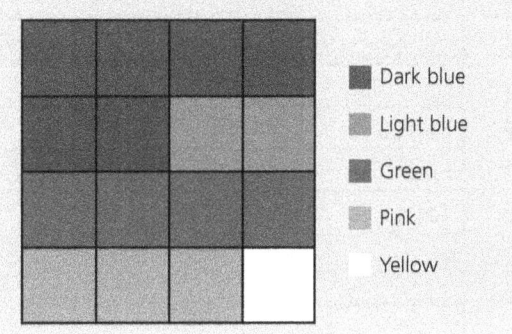

Dark blue
Light blue
Green
Pink
Yellow

Exercise 7.5 (page 52)

1 a Types of coffee sold as a percentage
b Students' answers
c Students' answers
d Students' own designs

2 a

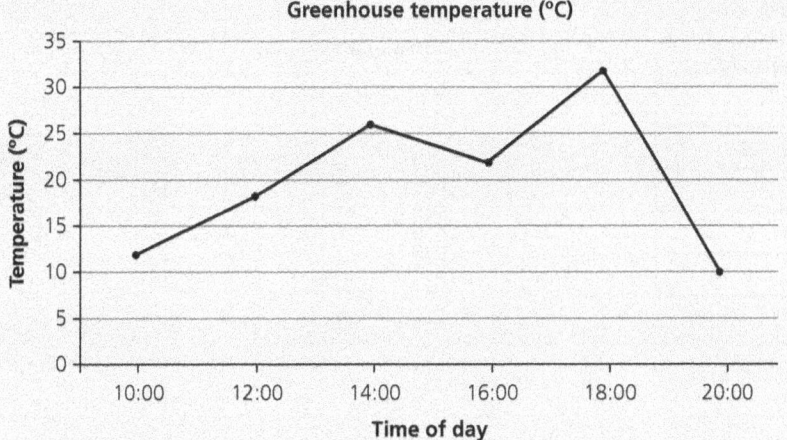

b Student circles temperature at 16:00 or 20:00
Temperature unlikely to dip in the middle of the day or temperature unlikely to dip so much between 18:00 and 20:00
c Student marks new position in line with existing data trend + explanation
d ≈ 22°C

3 a i C as temperature increases, ice-cream sales also likely to increase
ii B unlikely to be much of a correlation between the two subjects
iii A as a car gets older, its value tends to decrease
b Students' examples and justifications

Workbook answers

Exercises 7.1–7.2 (page 18)

1 a Ottawa
b 40
c Ottawa, Helsinki, London, Lisbon, Kathmandu, New Delhi, Bangkok
d 4

2

Weight (kg)	Tally	Frequency
2–3	II	2
3–4	III	3
4–5	III	3
5–6	I	1
6–7	III	3

3 a

	Cupcakes	Traybake	Large sponge	Total
Chocolate	75	12	**3**	90
Vanilla	**38**	**6**	10	**54**
Fruit	**42**	10	**2**	54
Total	155	**28**	15	**198**

b 38   **c** 28   **d** 36

4 a 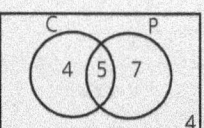   **b** 4   **c** 7

Exercises 7.3–7.4 (page 20)

1

Favourite colour	Frequency
Red	5
Blue	**4**
Green	3
Yellow	**2**
Other	6

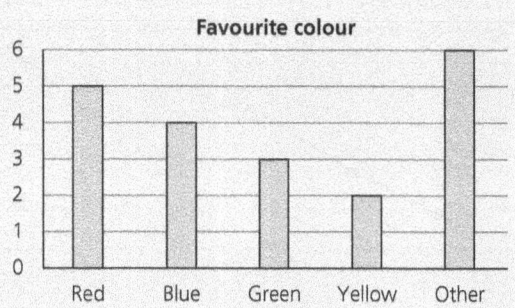

2 a

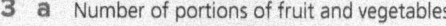

b £209

3 a Number of portions of fruit and vegetables

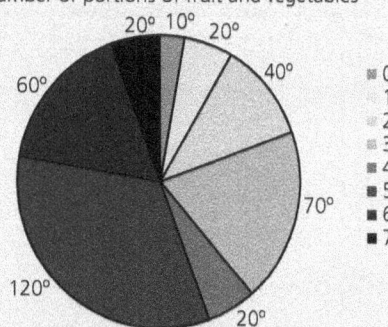

b

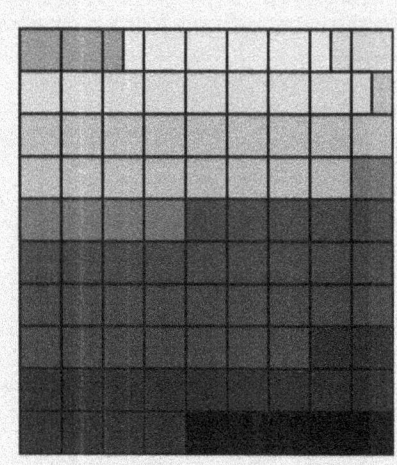

8 Properties of three-dimensional shapes

Prior knowledge

In prior learning, students will have had experience with a variety of 3D shapes. They will be able to recognise, name and sketch 3D shapes and their nets. Students will understand that the nets of 3D shapes are made from 2D shapes and will be able to associate this with the surface area of 3D shapes.

Objectives overview

Learning objective	Objective code	*Student's Book* pages	*Workbook* pages	*Teacher's Guide* pages	Online resources
Identify and describe the combination of properties that determine a specific 3D shape.	7Gg.06	54–63	22–26; 72–74	37–41	Flashcards Unit 8
Derive and use a formula for the volume of a cuboid. Use the formula to calculate the volume of compound shapes made from cuboids, in cubic metres (m³), cubic centimetres (cm³) and cubic millimetres (mm³).	7Gg.07	54–63	22–26	37–41	Knowledge test Unit 8 Worksheet: Euler's rule Worksheet: Surface area problem
Use knowledge of area and properties of cubes and cuboids to calculate their surface area.	7Gg.09	54–63	22–26	37–41	

Background information

In this unit, students will learn the names and sketch common 3D shapes. They will learn to identify a 3D shape by the number of faces, vertices and edges it has.

Students will also learn how to calculate the volume of some common 3D shapes. They will apply this knowledge to find the volume of compound 3D shapes.

Finally, students will learn how to find the surface area of a cuboid.

Terminology

It is important to use the correct mathematical terminology when describing shapes – students often refer to 'corners' when 'vertices' should be used, for example.

Lesson ideas

The *Student's Book* revises common 3D shapes. This would make a simple poster, which could be added to later. The exercise revises and extends basic ideas of 3D shapes.

Exercise 8.2 leads to a formula for volume of a cuboid without stating it yet.

To demonstrate composite shapes, building bricks such as Lego (or similar) are useful. Showing 2D to talk about 3D is not very good. Concrete 3D models are better especially for composite shapes. Students might have the bricks to make some for display (watch out for keeping them safe.)

Boxes which can be taken apart are good for work on surface area.

Starter activity

Wordsearch

Hidden in this grid are 12 words to do with 3D shapes.

Can you find all of them? Write down each word that you find.

S	P	W	S	E	X	T	S	K	A
O	N	H	V	E	R	T	E	X	U
M	F	O	P	Y	R	A	M	I	D
G	F	L	I	I	E	E	S	D	X
U	W	A	B	S	D	E	H	A	J
P	W	H	C	O	N	E	T	P	X
A	L	U	C	E	I	E	Y	T	S
C	B	C	V	O	L	U	M	E	G
E	D	G	E	Y	Y	K	O	I	Z
D	I	O	B	U	C	S	Q	U	D

Answers

CONE	CUBE	CUBOID
CYLINDER	DIMENSIONS	EDGE
FACE	NET	PYRAMID
SPHERE	VERTEX	VOLUME

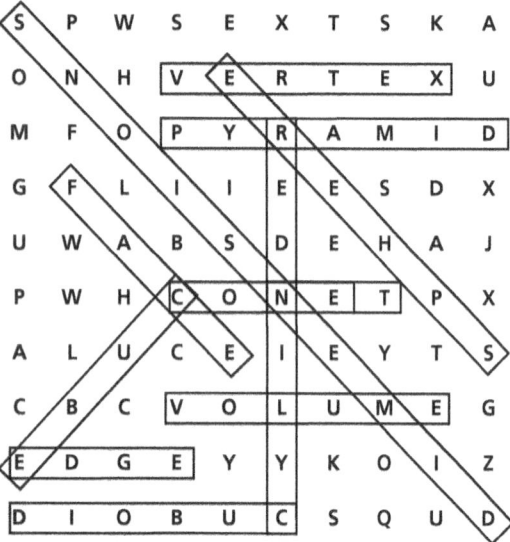

This activity is best delivered after students have learned about the surface area of cuboids and have completed Exercise 8.4 from their books.

Activity

Cube nets

Which of these nets can be folded to make a cube?

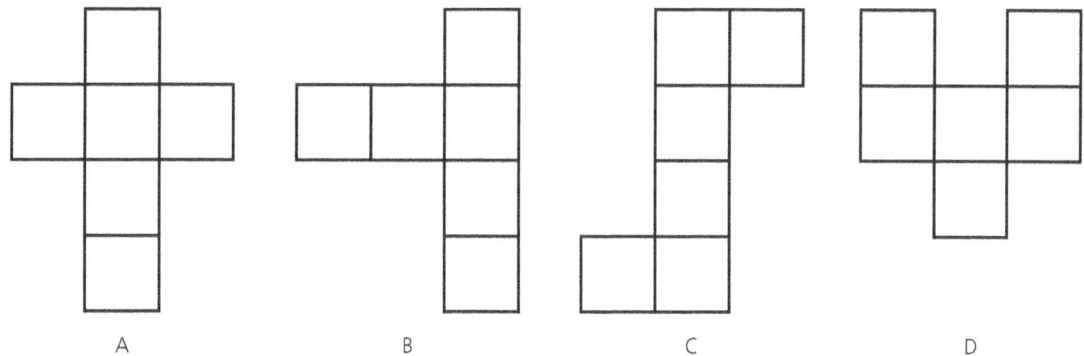

A   B   C   D

How many different nets for a cube can you find?

Hint: There are 11 different nets – can you find them all?

Answers

A and C can be made into a cube.

Student's Book answers

Exercise 8.1 (page 55)

1. a F=5; E=8; V=5
   b F=7; E=12; V=7
2. a Sphere   b Cone

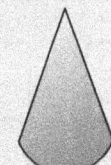

   c Triangular prism

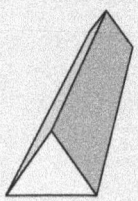

3. Students' examples, e.g. a cube and a pentagonal based pyramid
4. a i Yes, it is (6, 8, 12, 16)
      ii Students' explanations. Yes, it is always true as there are the same number of inclined edges as edges on the base of the pyramid, i.e. doubling the number of inclined edges will always give an even answer.
   b i Yes, it is (4, 5, 6 and 9)
      ii Students' explanations. Yes, because the number of vertices on the base = number of inclined faces. There is always one additional vertex, the apex and one additional face, the base.

Exercise 8.2 (page 57)

1.

	Length	Width	Height	Volume
a	2	2	2	8
b	2	8	1	16
c	2	4	3	24
d	4	6	1	24
e	4	4	4	64

Note: the numbers for each cuboid are interchangeable.

2. a length × width × height = volume
   b $V = l \times w \times h$
3. a 24 cm³   b 150 cm³
   c 40 cm³   d 40 000 cm³
   e 1500 cm³
4. 6 cm
5. 5 cm
6. a 8 cm
   b Students' answers will vary. The values must be different and their product must equal 64.
7. a $d = 12.8$ cm   b 10 in total

Exercise 8.3 (page 59)

1. 224 cm³
2. 1500 cm³
3. 384 cm³
4. Students' composite shape designs
5. Students' composite shape designs
6. 5.5 cm

Exercise 8.4 (page 62)

1. a 340 cm²   b 80 cm²
   c 128 cm²
2. a 150 cm²
   b 125 cm³
3. a Students' explanations
   b 600 cm²
4. a 54 cm²   b × 4
   c × 9
   d × 100 (scale factor)² = area factor
5. a Students' drawings of nets
   b 522 cm²
6. 4 pots
7. Fatou is correct. To prove this only one counter example is needed.

Workbook answers

Exercises 8.1–8.3 (page 22)

1

3D shape	Name	Number of faces	Number of edges	Number of vertices
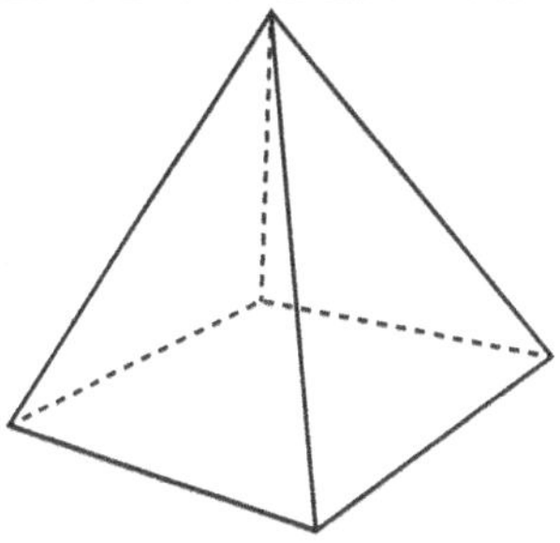	Square-based pyramid	5	8	5
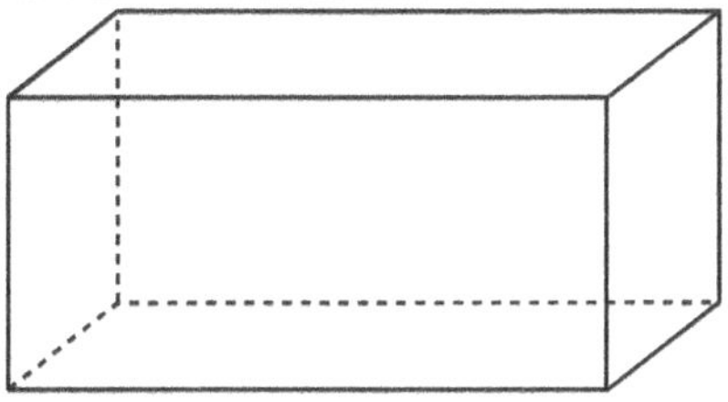	Cuboid	6	12	8
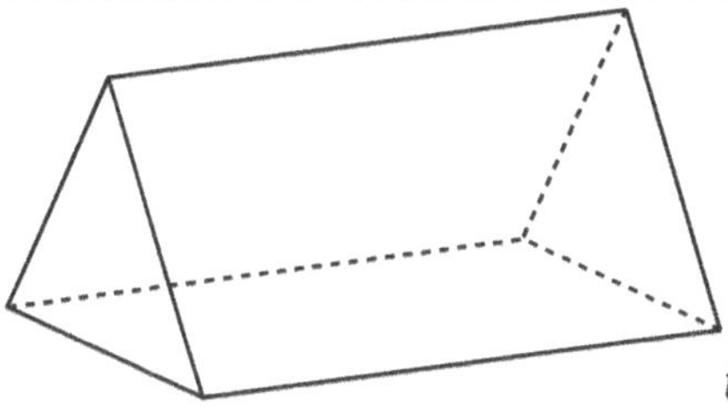	Triangular prism	5	9	6
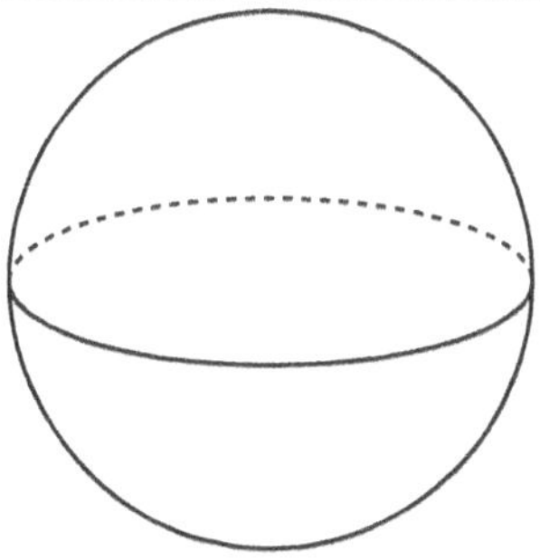	Sphere	0 (It has one surface, but not a face as it is not flat)	0	0

2 16 cm³

3 a Cuboid D   b 286 cm³

4 a 2 cm   b 4 mm   c 2.5 m

5 2 m

6 i, ii, iii 1, 1, 36   1, 2, 18   2, 2, 8

2, 3, 6   3, 3, 4

(And any other dimensions that multiply together to give 36)

7 320 litres

8 a 170 cm³   b 1656 cm³

Exercise 8.4 (page 25)

1 a 142 cm²

b 126 cm²

2 a 252 cm²

b You would need more plastic film as you would need an overlap.

9 Multiples and factors

Prior knowledge

In prior learning, students will have had experience of working with factors and multiples. They will know that 'common' factors and multiple are those shared by more than one number. Students will have used factors and multiples as a means of understanding divisibility tests.

Objectives overview

Learning objective	Objective code	*Student's Book* pages	*Workbook* pages	*Teacher's Guide* pages	Online resources
Understand lowest common multiple and highest common factor (numbers less than 100).	7Ni.04	64–69	27–28	42–44	Flashcards Unit 9

Knowledge test Unit 9

Worksheet: Prime sum |
| Use knowledge of test of divisibility to find factors of numbers greater than 100. | 7Ni.05 | 64–69 | 27–28 | 42–44 | |

Background information

In this unit, students will learn the terms 'highest common factors' (HCF) and 'lowest common multiples' (LCM) and find them from given sets of numbers.

Students learn the useful skill of being able to check whether a number is divisible by any of the numbers from 2 to 10 without having to use a calculator.

Terminology

Students often confuse factors and multiples. The shaded hundred squares in the *Student's Book* could be made into posters around the classroom to aid times table instant recall, whilst demonstrating what is meant by the two terms.

Lesson ideas

The *Student's Book* Unit 9 starts by revising factors and multiples and goes on to define LCM and HCF. Students will know about divisibility by 2, 5, 10 and with revision, about 3, 6, 9.

The unit assumes knowledge of spreadsheets. This might be done in another class in some schools. It is not essential.

Starter activity

Fizz buzz

Get into a group of four or more.

Each person takes it in turn to count on using the following rules:

1 When a number
   - is in the 2-times table say 'fizz'
   - is in the 3-times table say 'buzz'
   - is in the 2- AND 3-times table say 'fizz-buzz'

2 Otherwise, say the number as normal.

Support activity

Factor pairs

Factors come in pairs.

You can find factors by dividing your target number by 1, 2, 3, … and so on.

Example: Find all the factors of 20.

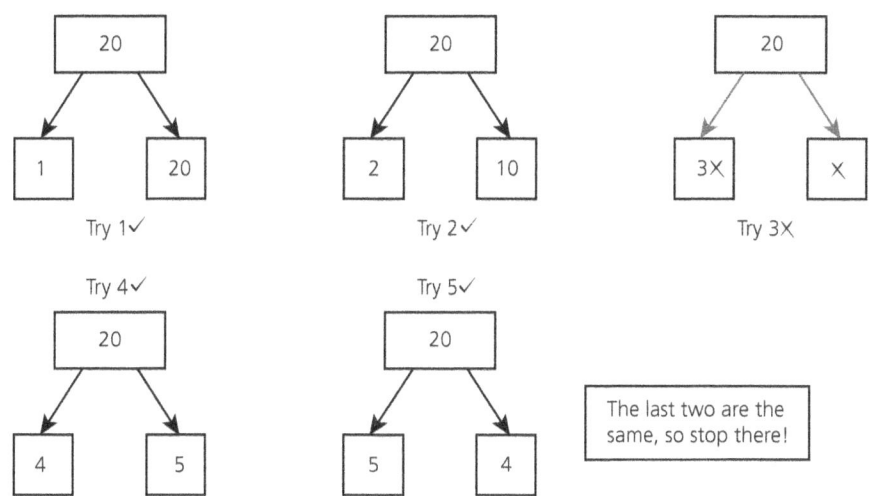

So, the factor pairs are: 1 and 20   2 and 10   4 and 5

The factors of 20 are: 1, 2, 4, 5, 10 and 20

Use this method to find all the factors of
   a 18   b 24   c 30   d 36   e 70   f 180

Answers
   a 1, 2, 3, 6, 9, 18   c 1, 2, 3, 5, 6, 10, 15, 30
   b 1, 2, 3, 4, 6, 8, 12, 24   d 1, 2, 3, 4, 6, 9, 12, 18, 36
   e 1, 2, 5, 7, 10, 14, 35, 70
   f 1, 2, 3, 4, 5, 6, 9, 10, 12, 15, 18, 20, 30, 36, 45, 60, 90, 180

Student's Book answers

Exercise 9.1 (page 65)

1 a 4   b 5   c 6   d 3   e 9

2 2

3 a 42   b 60   c 70   d 90   e 231

4 a One of 15 & 4 or 5 & 12 or 3 & 20 or 1 & 60.

   b One of 15 & 4 or 5 & 12 or 3 & 20 or 1 & 60, but not the same pair as used in part (a).

5 a Because in any other square, the total of its row or column is less than 24.

   b The three smallest factors of 24 are 1, 2 & 3 which gives a total greater than 4, therefore the blank must appear in the top row.

   c

1		3	4
4	6	12	22
24	8	2	34
29	14	17	

Exercise 9.2 (page 68)

1

	2	3	4	5	6	7	8	9	10	25	100
a 50	✓			✓					✓	✓	
b 270	✓	✓		✓	✓			✓	✓		
c 1120	✓		✓	✓		✓	✓		✓		
d 135		✓		✓				✓			
e 302 400	✓	✓	✓	✓	✓	✓	✓	✓	✓	✓	✓

2 a i   Any arrangement but last card must be 5.

     ii   Any arrangement as the sum of all four cards is divisible by 9.

     iii   Any arrangement as long as the last card is either a 2 or 8.

     iv   5328, 5832, 3528, 8352

   b Divisibility by 9 as the sum of all four cards is divisible by 9, therefore order does not matter.

3 a Students' own numbers tested for divisibility.

   b Students' own numbers tested for divisibility.

   c Students' own numbers tested for divisibility.

   d Yes because 9 is divisible by 3.

   e No. An example is the number 6, and 3 does not divide by 9 giving a whole number.

Workbook answers

Exercises 9.1–9.2 (page 27)

1 a 1, 32, 2, 16, 4, 8

   b 1, 100, 2, 50, 4, 25, 5, 20, 10

   c 1, 56, 2, 28, 4, 14, 7, 8

   d 1, 720, 2, 360, 3, 240, 4, 180, 5, 144, 6, 120, 8, 90, 9, 80, 10, 72, 12, 60, 15, 48, 16, 45, 18, 40, 20, 36, 24, 30

2 a 8, 16, 24, 32, 40, 48

   b 13, 26, 39, 52, 65, 78

   c 15, 30, 45, 60, 75, 90

   d 21, 42, 63, 84, 105, 126

3 a 7   b 48

   c 15 and 45   d 8 and 7

4 a 4   b 7

   c 6

5 a 30   b 70

   c 24

6 9.40 a.m.

7 5, 7, 8, 9

10 Probability and the likelihood of events

Prior knowledge

Students will know that probability is related to chance. They will be able to use the language of probability and will understand terms such as event and outcome. Students will be able to identify events that can happen simultaneously and will know what mutually exclusive events are.

Objectives overview

Learning objective	Objective code	Student's Book pages	Workbook pages	Teacher's Guide pages	Online resources
Use the language associated with probability and proportion to describe, compare, order and interpret the likelihood of outcomes.	7Sp.01	70–72	29–31	45–49	Flashcards Unit 10 Knowledge test Unit 10 End of Section 1 test
Understand and explain that probabilities range from 0 to 1, and can be represented as proper fractions, decimals and percentages.	7Sp.02	70–72	29–31	45–49	

Background information

In this unit, students learn how to describe chance using the correct mathematical terminology. They examine possible outcomes, their likelihoods and how to represent them as fraction, decimals or percentages.

 You can draw parallels between this and the use of language to express oneself in Cambridge Lower Secondary English, Speaking and Listening, Making yourself understood.

Terminology

The study of probability is language laden. As detailed in the *Student's Book*, we use words that are associated with probability all the time. For example, 'I *might* see that film', 'I'm *definitely* going to win this race' or 'It's *unlikely* that I'll pass this test'. Students might find it helpful to have a probability line as a classroom display so that probability language can be compared and linked to fractions, decimals and percentages.

Lesson ideas

The unit revises the vocabulary used in probability. Simple ideas of probability 1 or 0 can be discussed in pairs. Students should be familiar with simple Venn diagrams. (There may be one on display.)

Starter activity

 A dicey problem

Play this dice game with a friend:

Rules:

1 Choose a start position from A to F and place a counter on it.
2 Roll a dice. If you roll a:
   – 1 or 2: move 1 square left ←
   – 3, 4 or 5: move 2 squares right →
   – 6: move 1 square down ↓
3 If you move off the board to the left or right, you lose.
4 The winner is the first person to move their counter off the bottom of the board.

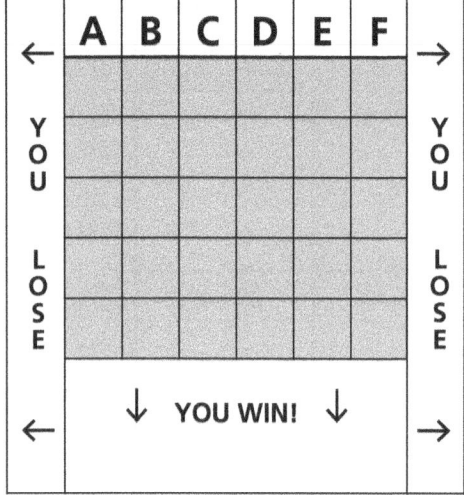

What is the best starting position?

TWM activity notes

Exercise 10.1 is explained here as an exemplar of Thinking and Working Mathematically (TWM), detailed in the Introduction to the *Teacher's Guide*, page xi. Students have covered how to calculate the theoretical probability of equally likely events and how this probability can be expressed as either a fraction, decimal or percentage.

Q3 Five cards are numbered with a different number from 1 to 10 as shown below. One card is covered.

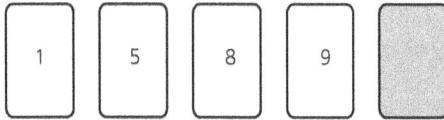

A card is chosen at random. What could be the number on the covered card if:
a the probability of picking an even number is 0.4
b the probability of picking a number less than five is 20%
c the probability of picking 10 is 1/5?

This is an example of a multi-step question in which students need to identify and then apply their grasp of different areas of mathematics in order to solve the problem. Looking at part (a) students will need to have an understanding of the relationship between decimals and fractions; the concept of random; equally likely events; and also, of course, what constitutes an even number.

TWM characteristics:　　　Specialising　Characterising　Classifying

This question if written in a standard way could have been presented as follows:

Q3　Five cards are numbered with a different number as shown below.

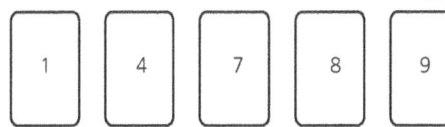

A card is chosen at random. Calculate:
a　the probability of picking an even number
b　the probability of picking a number less than five
c　the probability of picking 10.

Students are required to carry out simple theoretical probability calculations for random events.

Support activity
Probability

This reinforcement activity should be completed after all exercises in the student book have been explored.

1 An ordinary fair dice is rolled.
On the probability scale, mark the probability that the number rolled is

a	8	d	greater than 3
b	2	e	odd
c	less than 7	f	a factor of 12

2 A bag contains two red balls and three yellow balls.
A ball is chosen at random.
Write down the probability that the ball is

| a | red | b | yellow |
| c | red or yellow | d | green. |

3 Amir works out that the probability he is late to school is 110%.
Explain why Amir is wrong.

4 Jack and Chloe play a game.
They flip two coins.
Play the game a few times.
Is the game fair? Why/why not?

| 2 Heads Jack WINS | 2 Tails Jack WINS | 1 Head and 1 Tail Chloe WINS |

Answers

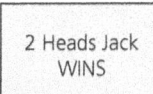

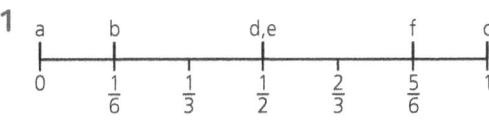

2 a 0.4　　　　　　　　　　　　　　　　**b** 0.6
　c 1　　　　　　　　　　　　　　　　　　**d** 0

3 A probability can't be more than 1 or 100% as a certain event has a probability of 1.

4 The game is fair: HH, TT is just as likely as TH, HT.

Student's Book answers

Exercise 10.1 (page 71)

1 a, b Students' words and probability scale

2 a i $\dfrac{1}{6}$ ii $\dfrac{1}{6}$

   iii $\dfrac{2}{6}$ or equivalent

  b i 10 times
    ii No, as experimental results may vary
      from theoretical probability.

3 a Either 2, 4, 6 or 10
  b Either 6, 7 or 10
  c 10

4 a TCA, TAC, CAT, CTA, ATC, ACT

  b i $\dfrac{1}{6}$ ii $\dfrac{2}{6}$ or equivalent

   iii $\dfrac{1}{2}$ iv $\dfrac{1}{6}$

5 a 80   b $\dfrac{28}{80} = \dfrac{7}{20} = 0.35 = 35\%$

6 a Yes, because as a ball is picked at
    random, then each number is equally
    likely to be picked.

  b i $\dfrac{1}{500}$ ii $\dfrac{1}{10}$

   iii 1   iv 0

Workbook answers

Exercise 10.1 (page 29)

1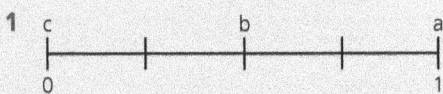

2 a 4
  b No, rolling a fair dice is random so there
    is no definite amount of 6s that will be
    rolled.
  c Cannot tell, the dice was not rolled
    enough times to say for sure if it is fair or
    biased.

3 $\dfrac{4}{9}$

4 a 1, 40, 2, 20, 4, 10, 5, 8

  b $\dfrac{5}{8}$

5 27 different combinations. 3×3×3.

6 a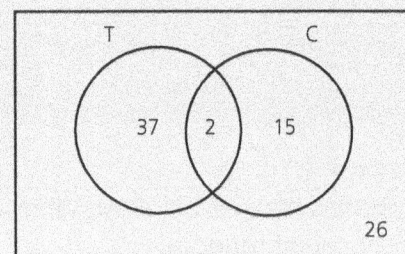

  b $\dfrac{59}{100}$ c $\dfrac{22}{37}$

Section 1 – Review

1 a

2	7	6
9	5	1
4	3	8

b Add 10 to each of the numbers in (a)

12	17	16
19	15	11
14	13	18

c i 45

ii Each number is the original number + 10. Each row, column, etc., has three numbers, therefore the total is the original total of $15 + 30 = 45$.

2

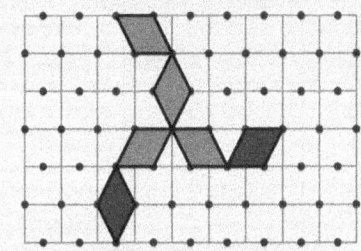

3 a Students' responses.
   b Students' responses.

4 $x = 8$ cm

5 a i He has added the 3 and the 5 first, but he should have multiplied the 5 by the 4 first.

ii 23

b $(3 + 5) \times 4$

6 $A = 52.5$ cm^2

7 Yes, they are showing the same results and the proportion shaded for each number is the same, i.e. $\frac{1}{8}$ shaded for 1, 4 and 5 people, $\frac{3}{8}$ shaded for 2 people and $\frac{1}{4}$ shaded for 3 people.

8 Side lengths P = 6 cm, Q = 3 cm

9 a i Any number.

ii For a number to be divisible by 4, only the last two digits must be divisible by 4. As 48 is divisible by 4, the missing number can be anything.

b i 3

ii For a number to be divisible by 9, the sum of the digits must be divisible by 9.
$3 + 3 + 4 + 8 = 18$ and 18 is divisible by 9

10 a $\frac{15}{30} = \frac{1}{2}$ **b** $\frac{5}{30} = \frac{1}{6}$

    c $\frac{1}{30}$ **d** $\frac{10}{30} = \frac{1}{3}$

11 Rounding and estimation – Calculations with decimals

 The introduction to Section 2 is about the roots of algebra. It can relate to Cambridge Lower Secondary English as students are encouraged to engage with the rich cultural roots of mathematics through the introduction.

Prior knowledge

In prior learning, students have explored place value. They will know the value of each digit in a decimal number and will be able to round figures with two decimal places to one decimal place or to a whole number. Students will be able to work with numbers with decimal places including estimation, addition and subtraction of decimals. Students have had prior experience of multiplication and division of decimals and whole numbers by 10, 100 or 1000.

Objectives overview

Learning objective	Objective code	Student's Book pages	Workbook pages	Teacher's Guide pages	Online resources
Round numbers to a given number of decimal places.	7Np.02	78–86	32–36	50–54	Flashcards Unit 11
Estimate, add and subtract positive and negative numbers with the same or different number of decimal places.	7Nf.07	78–86	32–36	50–54	Knowledge test Unit 11
Estimate, multiply and divide decimals by whole numbers.	7Nf.08	78–86	32–36	50–54	Worksheet: Squeeze
Use knowledge of place value to multiply and divide whole numbers and decimals by any positive power of 10.	7Np.01	78–86	32–36	50–54	

Background information

In this unit, students will learn how to round numbers to a given number of decimal places. They will examine this skill in a variety of contexts, including measures.

Being able to round numbers enables students to make estimates of answers to calculations when working with trickier or larger numbers. This 'self-checking' strategy is explored in this unit and will be of importance in their application of maths in many situations.

Students will also learn to multiply and divide by powers of 10 and continue to develop their written methods, extending into decimal numbers.

Terminology

There is a variety of vocabulary used to describe the power of a number. The power, also known as the index, tells you how many times you have to multiply the number by itself.

For example, 2^5 means that you have to multiply 2 by itself five times $= 2 \times 2 \times 2 \times 2 \times 2 = 32$.

10^2 is read as '10 to the power of 2' or '10 squared'. 10^3 is read as '10 to the power of 3' or '10 cubed'.

Lesson ideas

Students will have learned about decimals. It is worth revising decimal places and place value. The *Student's Book* does revise this too. Rounding a number from a calculator display is good practice. Students could do this in groups, setting their own questions and answering mentally.

The worked examples and explanation in the *Student's Book* are very clear and teachers can use them as a lesson plan. There are plenty of exercises and questions which get students to think and work mathematically in this unit. Estimation is emphasised and should be stressed as important by teachers.

Starter activity

 ### Estimation game

Play this estimation game with a friend.

The winner is the person who gave the best estimate in the most rounds.

Round 1: Just a minute!

You'll need one stopwatch each – or take it in turns.

Start a stopwatch and shut your eyes.

Open your eyes and stop the clock when you think 1 minute has passed.

The person closest to a minute, wins.

Round 2: How long?

Estimate the length of this line

_____

Measure the line to see who had the closest estimate.

Round 3: Spider senses!

Estimate the number of 🕷 in these rectangles.

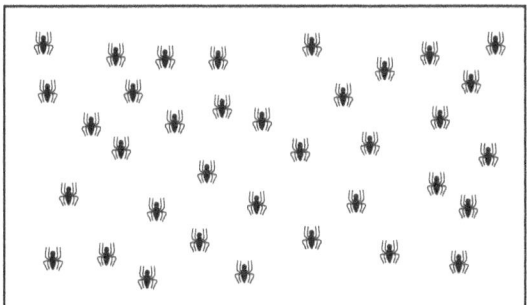 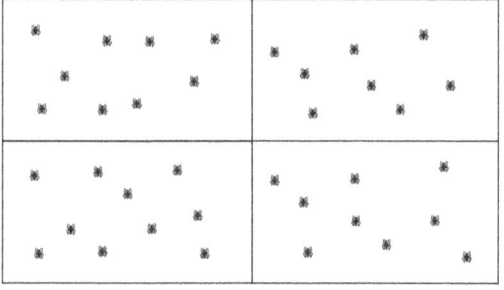

Count the spiders to see who was closest.

Who won?

What strategies did you use?

This activity is best suited for use once students have completed all exercises in Unit 11.

Support activity

Four-in-a-row

Play this game with a friend.

Choose a grid.

Take it in turns to choose a calculation from the list below and cross off the answer from your grid.

The winner is the first person to cross off a row or column of 4.

GRID A

0.12	1.7	3.4	0.8
3400	1.5	0.7	450
1	6.82	1.2	0.25
7	15	0.05	36
0.5	2.5	3.6	0.08

GRID B

0.08	1	2.5	3.6
450	1.7	36	7
0.8	0.12	0.7	3.4
4.5	0.5	15	0.25
1.2	1.5	3400	0.05

GRID C

1.7	1.5	0.5	0.08
2.5	0.05	0.12	0.7
0.8	3.6	3.4	7
0.5	0.25	3400	1
450	36	15	1.2

GRID D

7	1	0.5	3.6
0.7	36	0.25	3400
1.5	3.4	1.2	0.08
0.12	450	2.5	0.8
0.5	15	4.5	1.7

Calculation list

340×10	0.07×100	0.36×100
0.5×2	0.4×2	$2.4 \div 2$
0.9×5	0.02×4	0.45×100
$12 \div 100$	0.4×3	3.4×1000
$340 \div 100$	0.682×10	$2.5 \div 5$
12×0.3	$3400 \div 100$	$3.4 \div 2$
$34 \div 10$	3×0.5	$5 \div 10$
$682 \div 100$	0.015×1000	$5 \div 100$
$0.75 \div 3$	7×0.1	$70 \div 1000$
0.5×5	$250 \div 100$	0.1×7

Student's Book answers

Exercise 11.1 (page 79)

1 a 6.4   b 4.1   c 0.9   d 8.7

   e 1.1   f 0.1
2 a i 4.4   ii 4.38   b i 5.7   ii 5.72   c i 5.8   ii 5.80
   d i 1.5   ii 1.48   e i 4.0   ii 4.00   f i 6.3   ii 6.27
3 a i 0.57   ii 0.568   b i 3.48   ii 3.477   c i 8.85   ii 8.847
   d i 4.00   ii 4.000   e i 10.00   ii 10.000
4 Screen c
5 a $6.61   b $4.41
6 10.402s   10.395s   10.404s
7 14.069cm to 14.078cm

Exercise 11.2 (page 81)

1 a i Student's estimate   ii $116.72
   b i Student's estimate   ii $42.43
   c i Student's estimate   ii $111.06
   d i Student's estimate   ii $449.82
   e i Student's estimate   ii $91.26
2 a 83.6 kg   b 16.4 kg
3 18.6 cm
4 55 cm
5 a $53.25   b $6.75

Exercise 11.3 (page 82)

1 a 6300   b 46   c 8.4   d 0.65
   e 10.7
2 a 4500   b 720   c 96   d 4.85
   e 603.3
3 a 68   b 7.2   c 0.89   d 0.064
   e 0.0054
4 a 35   b 6.55   c 0.0562   d 0.008
   e 0.00034
5 a 46000   b 0.064   c 6800   d 0.046
   e 38000   f 840   g 700000   h 0.0095
   i 0.00000845   j 0.004
6
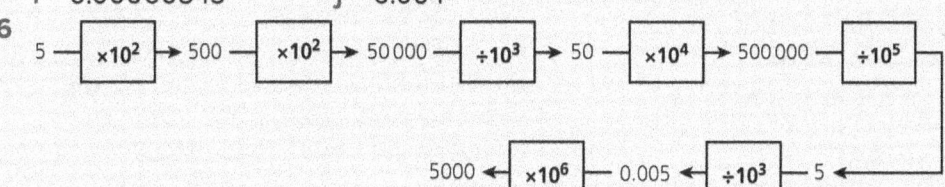

7 a Student's arrangement
   b Student's arrangement
   c No, order does not matter as multiplication and division can be done in any order
   d Student's arrangement e.g. $\div 10^2$ $\times 10^4$ $\div 10^3$ $\times 10^2$

Exercise 11.4 (page 86)

1 a i Student's estimate   ii 83.7
  b i Student's estimate   ii 96.6
  c i Student's estimate   ii 393.3
2 a i Student's estimate   ii 758.1
  b i Student's estimate   ii 154.28
  c i Student's estimate   ii 367.04
  d i Student's estimate   ii 167.91
3 b & d must definitely be wrong.
  b because multiplying 123.2 × 10 gives an answer bigger than the one given and 9.3 is less than 10.
  d because even dividing 127.8 by 2 would give an answer smaller than the one given.
4 $4071
5 a i Student's estimate   ii 3.5
  b i Student's estimate   ii 2.4
  c i Student's estimate   ii 5.5
  d i Student's estimate   ii 8.1
  e i Student's estimate   ii 3.3
  f i Student's estimate   ii 5.1
6 a 1.2   b 6.2   c 8.2   d 2.3
  e 4.9   f 1.09
7 a 0.5m   b 0.50m   c 0.503m

Workbook **answers**

Exercise 11.1 (page 32)

1 a 36.5   b 0.8
  c 465.73   d 45.22
  e 15.0   f 87.1
2 He has rounded it incorrectly. 12.36931688
  rounds to 12.4 to 1 decimal place.
3 36.724, 36.7163, 36.72, 36.718
4 a 6.3775cm
  b 6.3784cm

Exercise 11.2 (page 33)

1 a 173.122   b 86.09
2 a Yes, because 5+5+10+1+2 = 23
  b $23.38   c $6.62
3 a $35.25   b $34.08

Exercise 11.3 (page 35)

1 a 43   b 223700
  c 72500   d 82340

e 30.2   f 0.0437
g 3870000   h 0.042571
i 0.010043   j 0.0008362

2
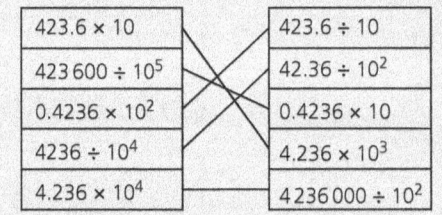

423.6 × 10		423.6 ÷ 10
$423\,600 \div 10^5$		$42.36 \div 10^2$
0.4236×10^2		0.4236×10
$4236 \div 10^4$		4.236×10^3
4.236×10^4		$4\,236\,000 \div 10^2$

Exercise 11.4 (page 36)

1 Students' answers close to, but not the
  same as, the following
  a 246373.95   b 116.4432
2 Students' answers close to, but not the
  same as, the following
  a 427   b 357.5
3 a 1856   b 32480
  c 18.56   d 324800

12 Mode, mean, median and range

Prior knowledge

In previous learning, students will have used the terms mean, median, mode and range. They will have worked with data sets and will be able to interpret them, identify patterns and compare data sets. Students will be able to make predictions and reach conclusions based on data.

Objectives overview

Learning objective	Objective code	Student's Book pages	Workbook pages	Teacher's Guide pages	Online resources
Use knowledge of median, mean, median and range to summarise large data sets.	7Ss.04	87–94	37–39	55–58	Flashcards Unit 12
Interpret data, identifying patterns, within and between data sets, to answer statistical questions. Discuss conclusions, considering the sources of variation, including sampling, and check predictions.	7Ss.05	87–94	37–39	55–58	Knowledge test Unit 12

Background information

The *Student's Book* unit begins by revising the terms mean, median and mode with worked examples. It then explains the term 'range' and encourages class discussion of these terms. The exercise makes these terms and their use clear.

 You can draw parallels between this and the use of language to express oneself in Cambridge Lower Secondary English, Speaking and Listening, Making yourself understood.

Terminology

Students will be able to define and calculate the different types of average given above. Please note that 'mode', 'modal value' and 'modal number' can be used interchangeably.

Lesson ideas

Frequency tables are taught in Unit 7 but are revised here with a worked example and further explanation. This unit is set out as a lesson plan.

Starter activity

 What are my numbers?

Solve these clues to find the numbers on the cards.

Each card has a **whole number** on it **from 1 to 10**

The cards are in **order of size**, smallest number first.

1

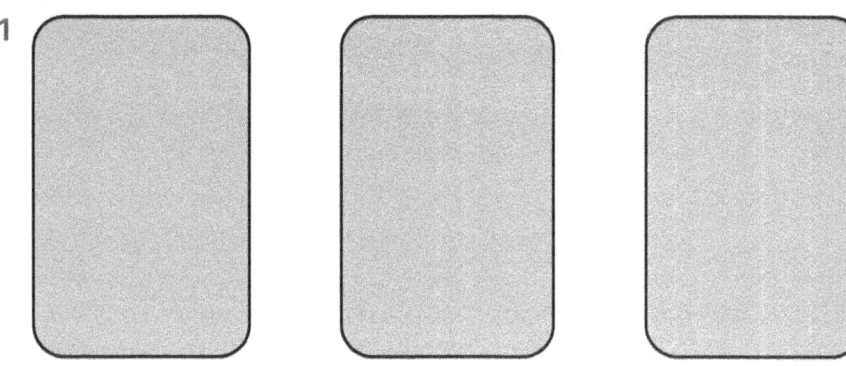

The middle card is 7.
The total of the cards is 20.
The difference between the highest and lowest cards is 7.

2

The middle card is 5.
The total of the cards is 28.
Only two of the cards are the same.
The difference between the highest and lowest cards is 7.
Most of the cards are odd numbers.
Can you find more than one answer?

3 Make up your own set of three or five numbers.
   Write clues for the numbers.
   Challenge a friend to work out your numbers.

Discussion Is it possible to have a 'middle' number when you have an even number of cards?

Answers

1 3, 7 and 10
2 3, 3, 5, 7, 10 or 2, 3, 5, 9, 9 or 2, 5, 5, 7, 9

TWM activity notes

Exercise 12.1 is explained here as an exemplar of Thinking and Working Mathematically (TWM), detailed in the Introduction to the *Teacher's Guide*, page xi. Students know how to calculate the mean, median, mode and range for a list of numbers and have been introduced to the idea that depending on the context, one value may be more useful than another.

Q2   Two discus throwers keep a record of their best throws (in metres) in the last ten competitions.

Discus thrower A	32	34	32	33	35	35	32	36	36	35
Discus thrower B	32	30	38	38	33	34	36	38	34	32

As a coach, you can only choose one of them for the next competition.
Which would you choose? Justify your choice mathematically.

Students are likely to calculate the mean, median, mode and range for this data as the unit deals with this.

The results are:

Discus thrower A:

mean = 34 m   median = 34.5 m   mode = 32, 35   range = 4 m

Discus thrower B:

mean = 34.5 m   median = 34 m   mode = 38   range = 8 m

There is no 'correct' answer to the question as the choice will be dependent on many factors. Students therefore will need to justify their choice. E.g. select A, because the mean and medians between the two discus throwers are very similar, but A is a more consistent performer due to the lower range value. Or choose B because although the mean and medians of the two discus throwers are very similar, if you want a better chance of getting a medal, B has shown that he is capable of throwing longer distances.

TWM characteristics: Convincing   Characterising   Critiquing

This question if written in a standard way could have been presented as follows:

Q2   Two discus throwers keep a record of their best throws (in metres) in the last ten competitions.

Discus thrower A	32	34	32	33	35	35	32	36	36	35
Discus thrower B	32	30	38	38	33	34	36	38	34	32

a *Calculate the mean, median, mode and range of the results for each discus thrower.*
b *In a competition, the coach wants to select the most reliable thrower. Which one does he choose?*
c *In a competition a gold medal will be awarded to a discus thrower who manages to throw a distance greater than 37 m. Which thrower should the coach select?*

Although these questions cover the same theory as the TWM style one, students have been given a more structured sequence of questions. In addition, there is no ambiguity relating to the possible answers as there is only one 'correct' answer to each.

Student's Book answers

Exercise 12.1 (page 90)

1　a　Mean = 3.27 (2 d.p.); median = 4; mode = 4; range = 4

　b　Mean = 6.36 (2 d.p.); median = 5.6; mode = 5.6; range = 5.9

　c　Mean = 2.92 (2 d.p.); median = 3; mode = 4; range = 6

2　Student's choice and justification

3　a, b, c　Student's data collection and interpretation

4　a　Mean = 9.6; median = 10; mode = 10

　b　Only the mean will be affected. The median is unaffected as both cards with 10 will be the middle pair. The mode is unaffected as there will still be more of the number 10 than any other number.

　c　The mean will definitely be affected. The median too will also be affected as the middle pair will now be the 7 and 10, so the median will be 8.5. The mode **may** change if the sixth card is a 6 then there will be two modes, 6 and 10. If it's less than 6, then the mode will still be 10.

5　2006 drove off.
1971+1986+1993+2006+2010 = 9966
4×1990 give the sum of the years of the remaining four cars = 7960
9966 − 7960 = 2006

6　88.4 kg

7　94 points

8　Yes, as the mean weight of the 20 cans is 411.4 g. Students may answer no and give the justification that the difference is so small that it may be decided to leave the machine as it is.

9　a　i　Student's list of numbers
　　　ii　Student's explanation
　b　i　Student's list
　　　ii　Student's explanation
　c　i　Student's list
　　　ii　Student's explanation

10　a　i　Student's list (e.g. 2, 2, 3, 4, 10)
　　　ii　Not possible
　　　iii　Student's list (e.g. 1, 1, 10, 11, 12)
　b　Student's explanation of why ii) is not possible.

Exercise 12.2 (page 93)

1　1st chapter as it is the biggest sample size.

2　a　i　Mean = 2.78 (2 d.p.)
　　　ii　Mode = 3
　　　iii　Median = 3
　　　iv　Range = 6
　b　Student's assumption, e.g. Competitors continue to catch fish at a constant rate, Fish don't run out etc.

3　a　10
　b　10
　c　3 & 4
　d　3
　e　No, because weekend results are likely to differ from weekday results.

Workbook answers

Exercises 12.1–12.2 (page 37)

1　a　Mean = 5; mode = 3 and 5; median = 5; range = 6

　b　Mean = 6.05; mode = no mode; median = 6; range = 7.8

　c　Mean = 46; mode = 30; median = 45; range = 40

2　a　6.5m is an outlier and would skew the data.

　b　Median, it would rule out any outliers. There is no mode so this would not be useful.

3　a　1, 1, 6, 6, 6　　　　b　5, 7, 8, 15, 15

4　4

5　$10

6　8

7　22

Transformations of two-dimensional shapes

Prior knowledge

Students will have had previous experience working with 2D shapes. They will be able to reflect 2D shapes in a mirror line and rotate 2D shapes 90 degrees. Students will also be able to identify rotational symmetry and complete symetrical patterns.

Objectives overview

Learning objective	Objective code	*Student's Book* pages	*Workbook* pages	*Teacher's Guide* pages	Online resources
Identify reflective symmetry and order of rotational symmetry of 2D shapes and patterns.	7Gg.10	95–108	5–6; 40–44	59–67	Flashcards Unit 13 Knowledge test Unit 13 Worksheet: Triangles and shapes
Reflect 2D shapes on coordinate grids, in a given mirror line (*x*- or *y*-axis), recognising that the image is congruent to the object after a reflection.	7Gp.04	95–108	40–44	59–67	
Rotate shapes 90° and 180° around a centre of rotation, recognising that the image is congruent to the object after a rotation.	7Gp.05	95–108	40–44	59–67	
Understand that the image is mathematically similar to the object after enlargement. Use positive integer scale factors to perform and identify enlargements.	7Gp.06	95–108	40–44	59–67	

Background information

In this unit, students begin by revising reflective and rotational symmetry. They extend this knowledge by reflecting and rotating on coordinate axes. Students then examine enlargement as an additional type of transformation.

Terminology

Reflection, rotation, enlargement and translation are all forms of transformation.

Students should learn the appropriate vocabulary to use in their study of transformations, for example, when describing the rotational symmetry of a shape, to use 'the order of …' to describe the number of times the shape looks the same in one complete rotation.

Students are introduced to the concept of congruency. Shapes which are congruent are exactly the same size and shape. They can be reflections and/or rotations of each other. Their interior angles are equal.

In the starter activity, students are introduced to tessellation in the context of tiling patterns.

Lesson ideas

When revising their understanding of reflective symmetry and rotational symmetry, students may prepare a poster to show this. There is a discussion point in 'Let's talk' which is very useful for clarifying terms.

The big table of shapes and symmetry is perfect for a large display poster. There are further exercises on reflection and rotation. There are further examples and explanation on Translation and Enlargement with examples and exercises.

This unit is set out as a detailed lesson plan for teachers.

Starter activity

Escher tessellations [suitable flipped learning task]

A tessellation is a tiling pattern. The tiles completely cover a space leaving no gaps or overlaps.

M.C. Escher (1898–1972) was a Dutch artist who produced amazing tessellations. Look up his artworks.

Follow these steps to produce your own Escher style tessellation.

Step 1: Take a piece of card.

Step 2: Draw a wavy line down the right-hand side.

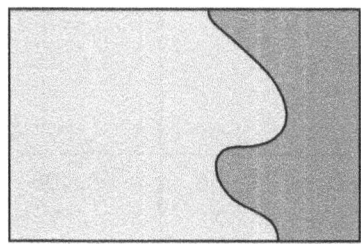

Step 3: Cut out along your wavy line. Place your cut out on the left-hand side.

Step 4: Draw a triangle at the top of your card.

Step 5: Cut out along your triangle. Place your cut out at the bottom.

Step 6: Now you have a shape that tessellates. Decorate your shape and make a tessellation pattern.

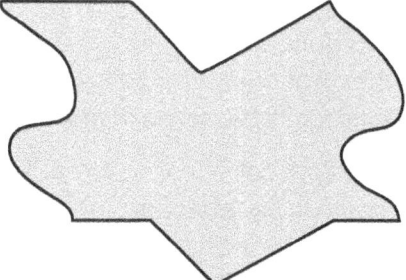

Student's Book answers

Exercise 13.1 (page 96)

1

Shape		Number of lines of reflective symmetry	Order of rotational symmetry
Square		4	4
Rectangle		2	2
Rhombus	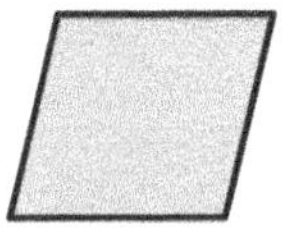	2	2
Parallelogram	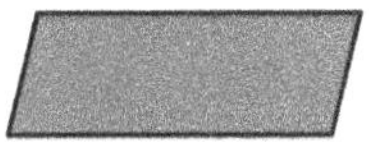	0	2
Isosceles trapezium		1	1
Kite	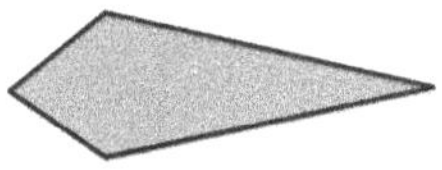	1	1
Equilateral triangle	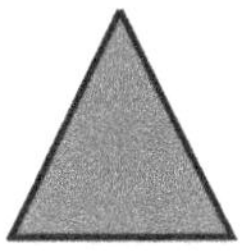	3	3
Isosceles triangle	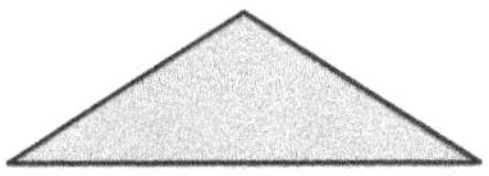	1	1
Regular pentagon	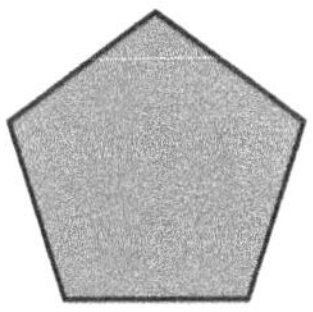	5	5
Regular hexagon	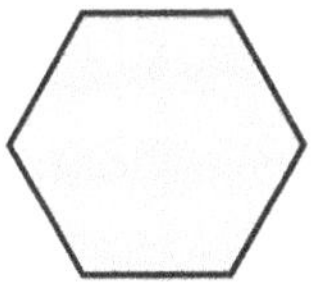	6	6
Circle		Infinite	Infinite

2 a i 1   b i 1   c i 1   d i 2
   ii 1   ii 1   ii 1   ii 2
  e i 2   f i 0   g i 0   h i 0
   ii 2   ii 2   ii 2   ii 2

3 Other answers are possible.

a   b   c

d   e   f

4 Other solutions are possible.

a   b   c

d

5 a   b

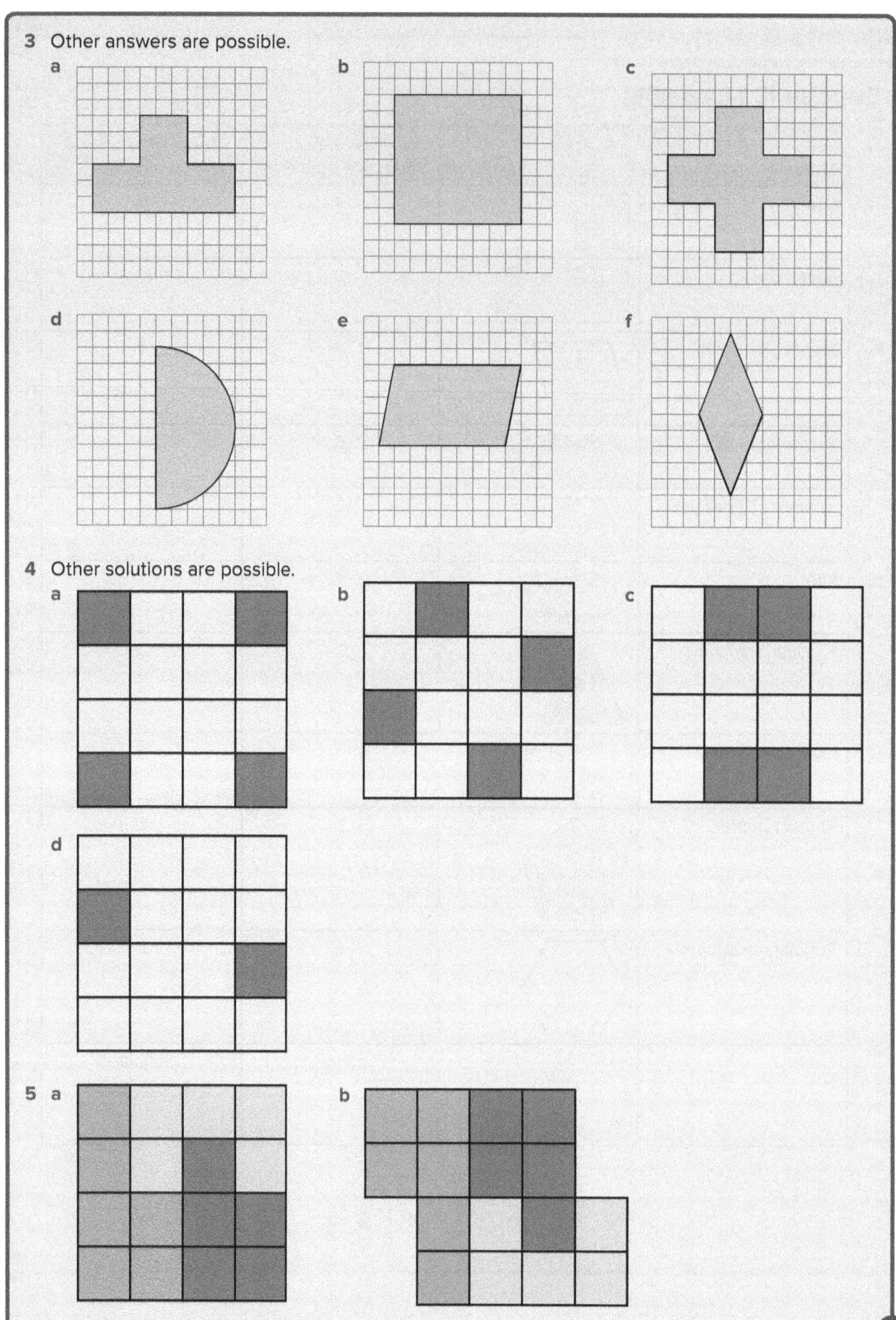

Exercise 13.2 (page 101)

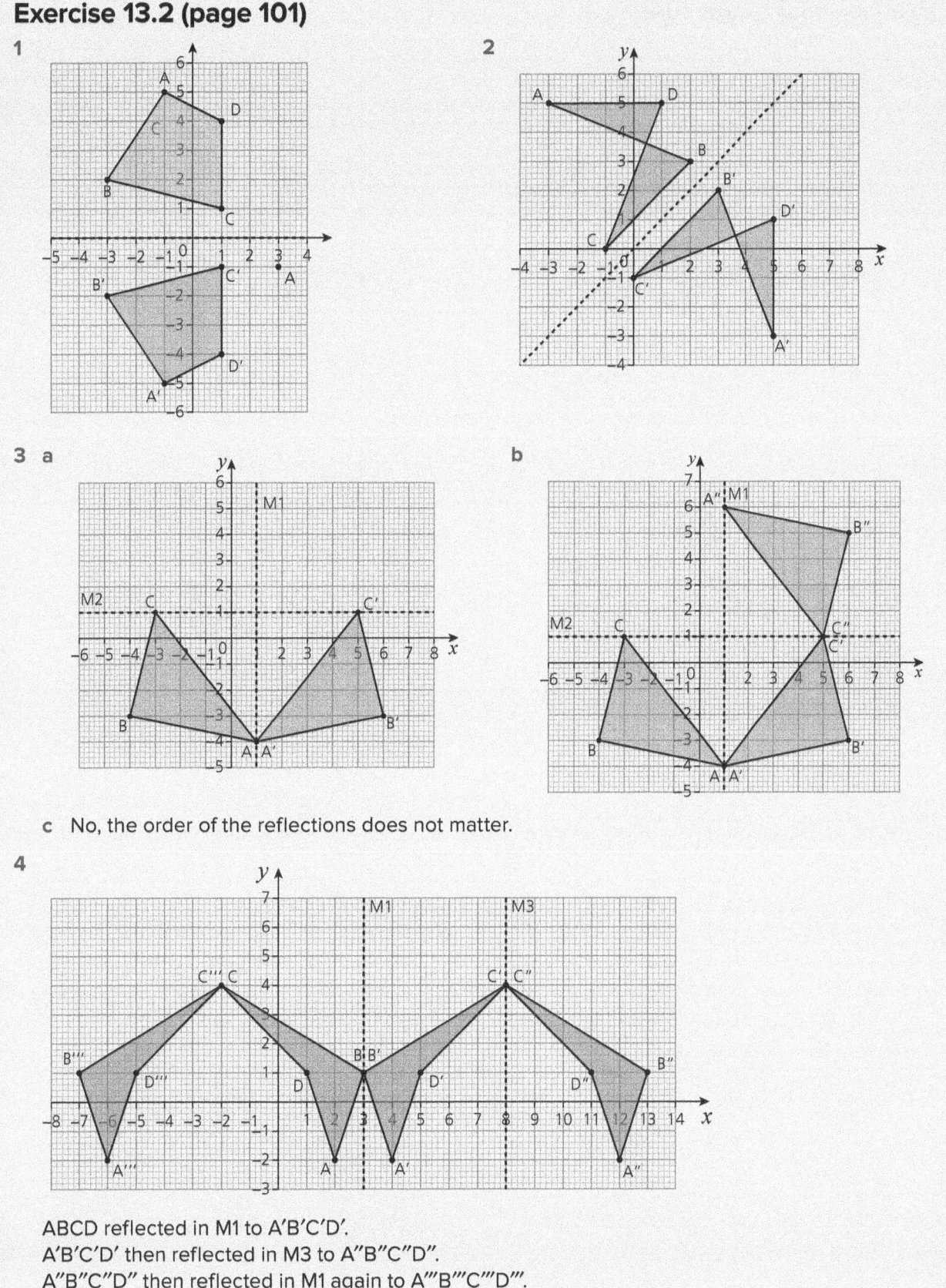

c No, the order of the reflections does not matter.

ABCD reflected in M1 to A'B'C'D'.
A'B'C'D' then reflected in M3 to A"B"C"D".
A"B"C"D" then reflected in M1 again to A'''B'''C'''D'''.

Exercise 13.3 (page 104)

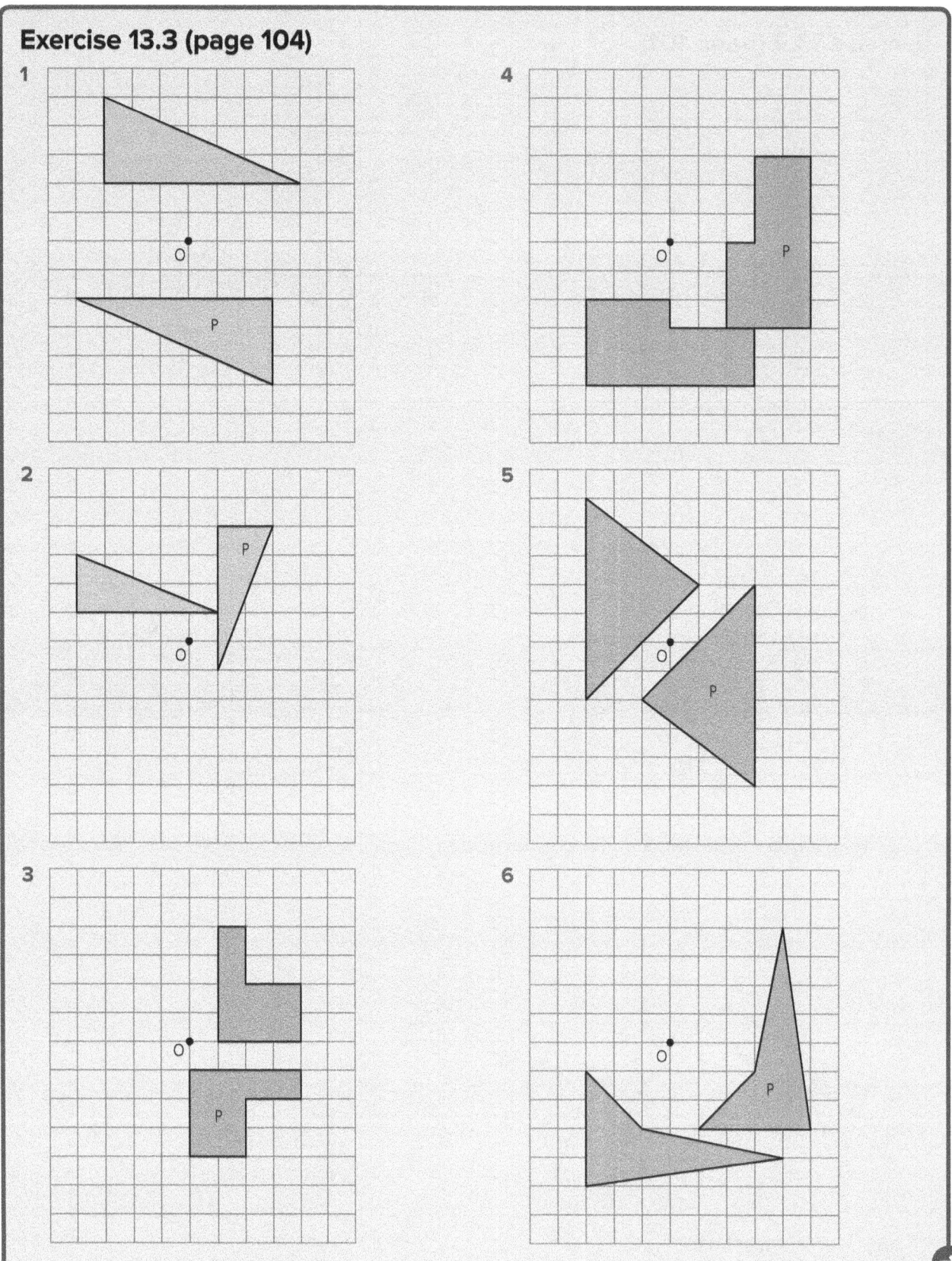

7   Both could be correct with the information given on the diagram.
    The vertices of the squares need to be labelled (e.g. ABCD) to see where they have been
    mapped to after the transformation.

Exercise 13.4 (page 107)

1   a   Yes, B is an enlargement of A.   b   Scale factor of enlargement is 2.
2   a   No, B is not an enlargement of A as the length has been doubled but not the height.
3   a   Yes, B is an enlargement of A.   b   Scale factor of enlargement is 4.
4   a   Yes, B is an enlargement of A.   b   Scale factor of enlargement is 2.
5   a   No, B is not an enlargement of A as the width has doubled but the height has trebled.
6   a   No, B is not an enlargement of A as they are congruent shapes and for an enlargement the
        scale factor ≠ 1.

Exercise 13.5 (page 108)

The position of the enlarged shape may vary from those shown.

1
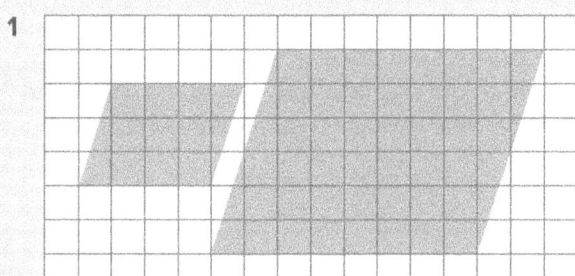

2
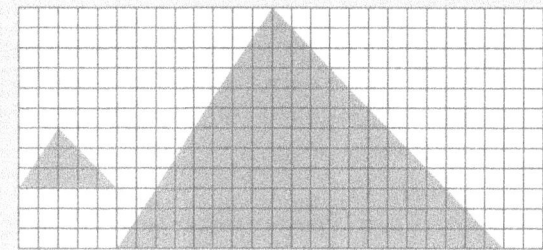

3
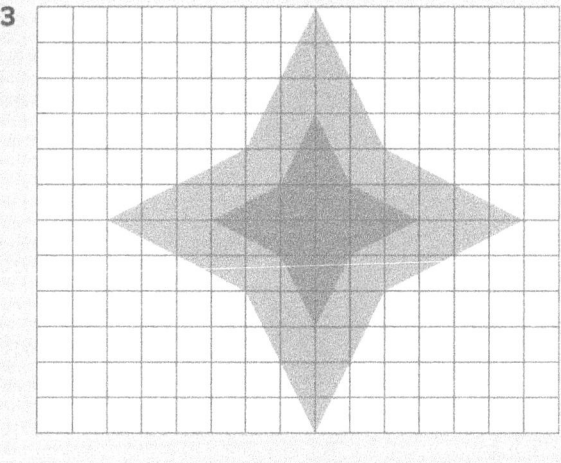

4
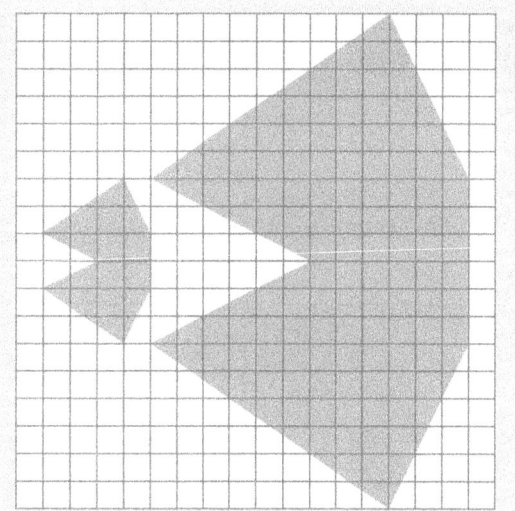

5

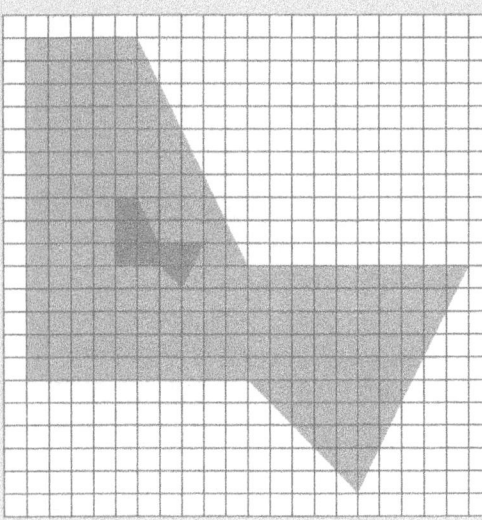

6

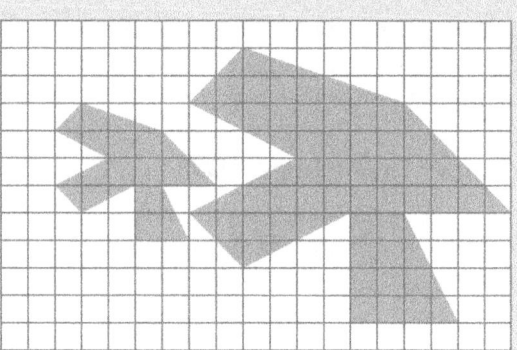

7 a R′ is incorrect.
R and S are vertically in line with each other. R′ and S′ must be too, therefore, one of them must be incorrect. To get from Q to R the coordinates are +5 in the x-direction and −2 in the y-direction. As the enlargement is a scale factor of 2, to get from Q′ to R′ should be +10 in the x-direction and −4 in the y-direction but this is not the case as it is +9 in the x-direction and −4 in the y-direction.

b R′ (13, 2)

Workbook answers

Exercises 13.1–13.2 (page 40)

1 a i

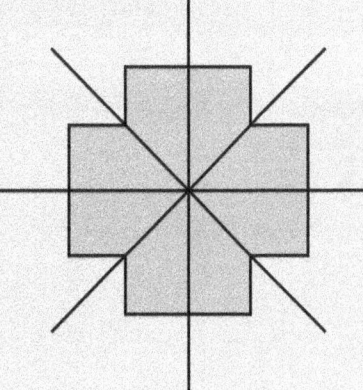

ii Rotational symmetry order 4.

b i

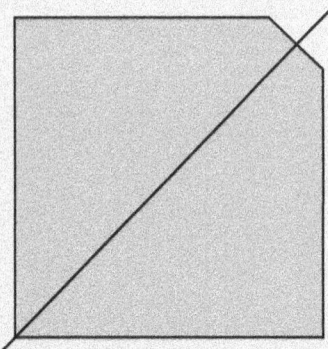

ii Rotational symmetry order 1.

2 These are an example of answers; students may have other examples.

a Square

b Parallelogram

c   Isosceles trapezium (or kite)

3

4 a

b

5   a, b, c

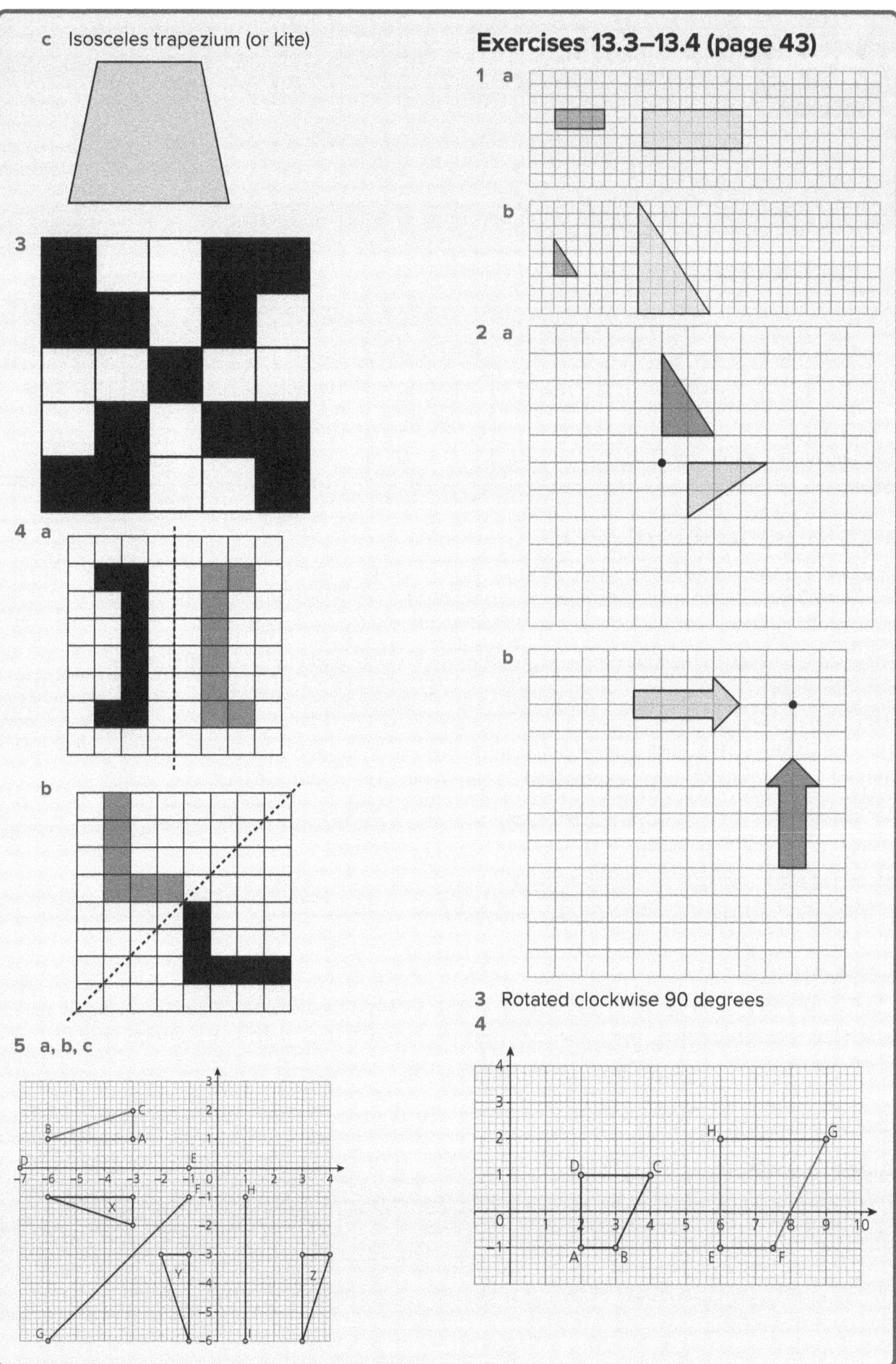

Exercises 13.3–13.4 (page 43)

1 a

b

2 a

b

3   Rotated clockwise 90 degrees
4

14 Manipulating algebraic expressions

Prior knowledge

Students should revise and understand the work done in algebra in Section 1.

Objectives overview

Learning objective	Objective code	*Student's Book* pages	*Workbook* pages	*Teacher's Guide* pages	Online resources
Understand how to manipulate algebraic expressions including: collecting like terms; applying the distributive law with a constant.	7Ae.03	28–35; 109–115	14–17; 45–46; 57–59	68–70	Flashcards Unit 14 Knowledge test Unit 14

Background information

In this unit, students will learn how to simplify algebraic expressions by combining like terms. There are many occasions when an algebraic expression can be simplified. Students are taught that it is good mathematical practice to leave expressions written in their simplest form.

They will learn how to multiply simple expressions and how to expand those with brackets.

Terminology

This unit begins with a quick revision of the term 'expression' and some worked examples. The student book explains that an **expression** is used to represent a value in algebraic form.

For example,

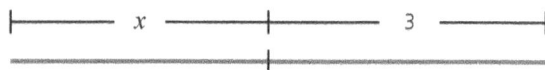

The length of the line is given by the expression $x+3$.

Lesson ideas

Exercise 14.1 teaches simplifying expressions. The unit then shows how to expand an expression which has brackets followed by an exercise. The final question deals with composite shapes.

The unit is set out as a detailed lesson plan for teachers.

Starter activity

 Algebra walls

The expression in each brick in these walls is found by adding the expressions in the two bricks beneath it. Fill in the missing expressions in each brick.

1

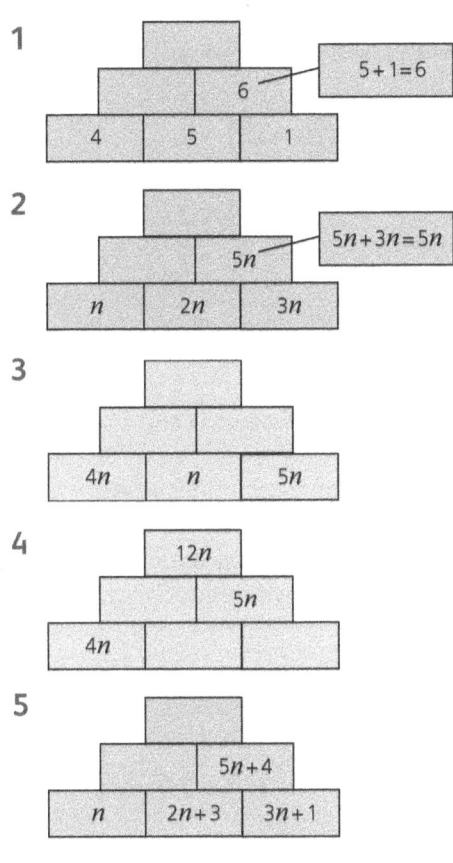

Bottom row: 4, 5, 1. Middle: 6 ← 5 + 1 = 6.

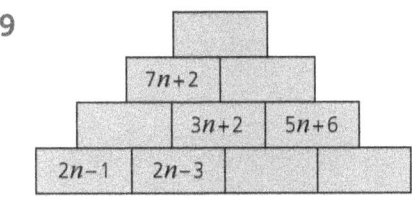

2

Bottom row: n, $2n$, $3n$. Middle: $5n$ ← $5n + 3n = 5n$

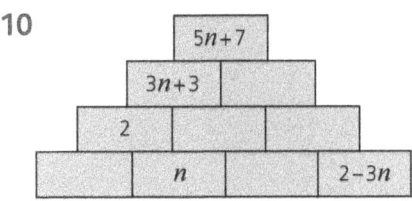

3

Bottom row: $4n$, n, $5n$.

4

Top: $12n$. Middle: (blank), $5n$. Bottom: $4n$, (blank), (blank).

5

Middle: (blank), $5n + 4$. Bottom: n, $2n + 3$, $3n + 1$.

6

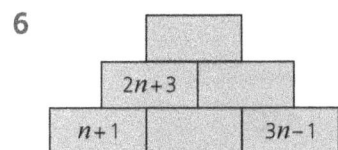

Middle: $2n + 3$. Bottom: $n + 1$, (blank), $3n - 1$.

7

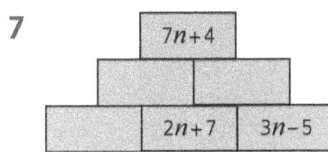

Top: $7n + 4$. Bottom: (blank), $2n + 7$, $3n - 5$.

8

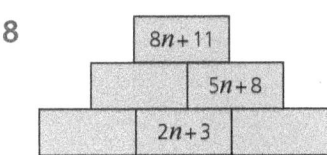

Top: $8n + 11$. Middle: $5n + 8$. Bottom: $2n + 3$.

9

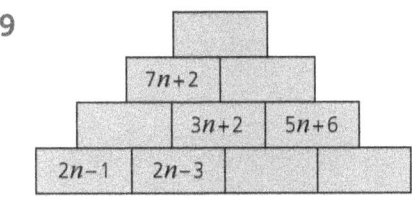

Middle: $7n + 2$. Next: $3n + 2$, $5n + 6$. Bottom: $2n - 1$, $2n - 3$.

10

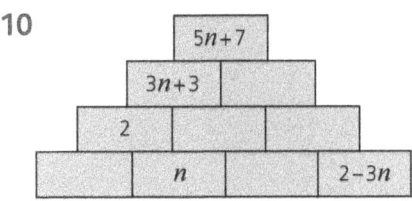

Top: $5n + 7$. Next: $3n + 3$. Next: 2. Bottom: n, $2 - 3n$.

Answers

1

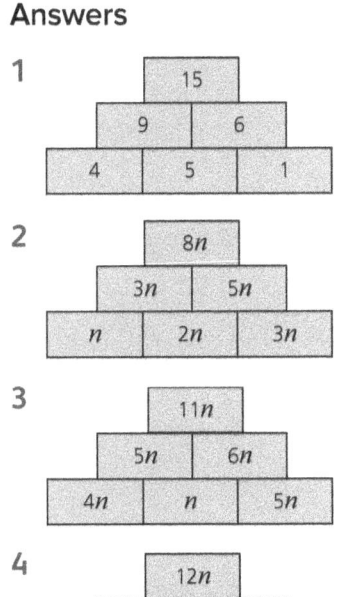

Top: 15. Middle: 9, 6. Bottom: 4, 5, 1.

2

Top: $8n$. Middle: $3n$, $5n$. Bottom: n, $2n$, $3n$.

3

Top: $11n$. Middle: $5n$, $6n$. Bottom: $4n$, n, $5n$.

4

Top: $12n$. Middle: $7n$, $5n$. Bottom: $4n$, $3n$, $2n$.

5

Top: $8n + 7$. Middle: $3n + 3$, $5n + 4$. Bottom: n, $2n + 3$, $3n + 1$.

6

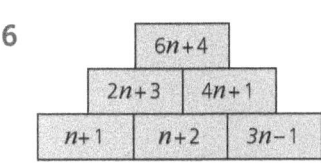

Top: $6n + 4$. Middle: $2n + 3$, $4n + 1$. Bottom: $n + 1$, $n + 2$, $3n - 1$.

7

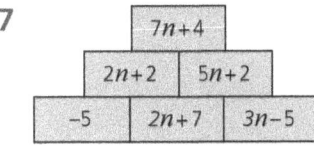

Top: $7n + 4$. Middle: $2n + 2$, $5n + 2$. Bottom: -5, $2n + 7$, $3n - 5$.

8

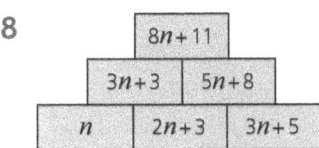

Top: $8n + 11$. Middle: $3n + 3$, $5n + 8$. Bottom: n, $2n + 3$, $3n + 5$.

9

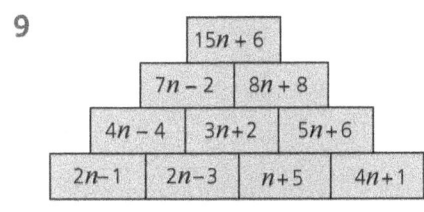

Top: $15n + 6$. Next: $7n - 2$, $8n + 8$. Next: $4n - 4$, $3n + 2$, $5n + 6$. Bottom: $2n - 1$, $2n - 3$, $n + 5$, $4n + 1$.

10

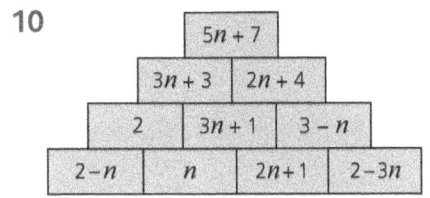

Top: $5n + 7$. Next: $3n + 3$, $2n + 4$. Next: 2, $3n + 1$, $3 - n$. Bottom: $2 - n$, n, $2n + 1$, $2 - 3n$.

Student's Book answers

Exercise 14.1 (page 110)

1 a $10a$ b $9b$
   c $6c+15$ d $2d+3$
   e $10e+7$ f $3f-11$
   g $-4g-3$ h $-6h+4$
   i $12i-18$ j $10j-15$

2 a $2l+2w$ b $6b+12$
   c $3s+6$ d $4x+2y+2$
   e $10m+6n$

3 a $16p$
   b i $4p$
     ii The perimeter is unchanged as
       $5p+3p+4p+p+p+2p$ simplifies to
       $16p$.

4 a i all except $2x+1$
     ii Other arrangements are possible.

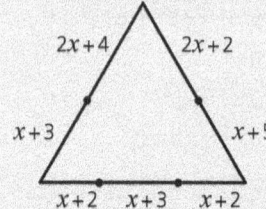

   b $3x+7$

Exercise 14.2 (page 113)

1 a $4a+12$ b $3b-6$
   c $15c-5$ d $2d-12$

2 10 units

3 a $14j+15$ b $7h+20$

4 $40x-20$

5 a A or B as the rectangle must have a
    length of $2x+1$ and width of 2 which
    would produce the area of A and the
    perimeter of B.
   b Not necessarily as a rectangle with
    an area of $4x+2$ units² could be
    formed from a rectangle with different
    dimensions from those needed, e.g.
    $4(x+\frac{1}{2})$. A perimeter of $4x+6$ units
    could be formed from a rectangle
    different from that needed, e.g. length
    $2x$ and width 3 units.

Workbook answers

Exercises 14.1–14.2 (page 45)

1 a $14a$ b $12b$
   c $8c+8$ d $5d+3$
   e $15e+3$ f $9f-4$

2 a $2p+12$ b $5d+5$
   c $26f+30$

3 a $8a+12$ b $6c-30$
   c $6d+2cd$ d $6f-12g+30$

4 a $30g+40$ b $20g+24$
   c $15g+20$

5 16

Fractions, decimals and percentages

Prior knowledge

Students have had experience working with fractions in prior learning. They will be able to estimate, multiply and divide fractions by whole numbers and write fractions in their simplest form. Students will know that fractions, decimals and percentages have equivalent values.

Objectives overview

Learning objective	Objective code	*Student's Book* pages	*Workbook* pages	*Teacher's Guide* pages	Online resources
Recognise that fractions, terminating decimals and percentages have equivalent values.	7Nf.01	116–129	47–50; 69–71	71–77	Flashcards Unit 15
Estimate and add mixed numbers, and write the answer as a mixed number in its simplest form.	7Nf.02	116–129	47–50	71–77	Knowledge test Unit 15
Estimate, multiply and divide proper fractions.	7Nf.03	116–129	47–50	71–77	Worksheet: Fraction triangle
Use knowledge of common factors, laws of arithmetic and order of operations to simplify calculations containing decimals or fractions.	7Nf.04	116–129	47–50	71–77	
Understand the relative size of quantities to compare and order decimals and fractions using the symbols =, ≠, > and <.	7Nf.06	116–129	47–50	71–77	

Background information

In this unit, students learn how to convert numbers between their fraction, decimal and percentage equivalents. They revise how to add and subtract fractions with the same (common) denominator and extend to fractions with differing denominators, using their understanding of finding lowest common multiples.

The unit then introduces students to mixed numbers and improper fractions, before examining the multiplication and division of fractions, considering the use of common factors to simplify calculations where appropriate.

Terminology

This unit introduces the term 'reciprocal'. The final section says to use the 'commutative' property of multiplication to make it easier by multiplying in any order.

Lesson ideas

The unit begins with revision of the equivalence of fractions, decimals and percentages. The table in Exercise 15.1 would make a good poster.

There is an opportunity for group discussion before the last questions which help students to think and work mathematically.

Starter activity

Shady shapes

1 Write down the proportion of each shape that is shaded.
Give your answers as a percentage, fraction and decimal.

Percentage shaded = _____
Fraction shaded = _____
As a decimal, the proportion shaded is _____

Percentage shaded = _____
Fraction shaded = _____
As a decimal, the proportion shaded is _____

2 Find an interesting way to shade an exact fraction of these grids.

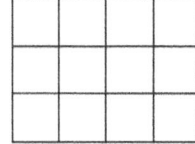 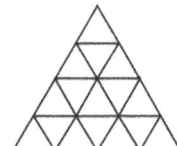

What fractions did you shade?
Write each of your fractions as a decimal and a percentage.

Answers

1 percentage shaded = 25%, fraction shaded = $\frac{1}{4}$, decimal shaded = 0.25

percentage shaded = 50%, fraction shaded = $\frac{1}{2}$, decimal shaded = 0.50

2 Students' own answers.

TWM activity notes

Exercise 15.5 is explained here as an exemplar of Thinking and Working Mathematically (TWM), detailed in the Introduction to the *Teacher's Guide*, page xi. Students will already have covered how to multiply decimals by powers of 10, in addition to the use of mental strategies for simplifying calculations. Calculators are discouraged for this question.

Q1 a *The rectangle below has the dimensions shown.*

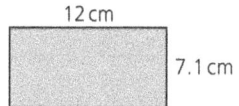

12 cm

7.1 cm

Explain why the area of the rectangle can be calculated by the calculation
$(7.1 \times 10) + (7.1 \times 2)$

b *Work out the area of the rectangle below.*

15.3 cm

8 cm

An important component of TWM is for students to articulate their mathematical thinking and mathematical processes. Part (a) of this question provides a framework to assist students in this process. With part (b), splitting the shape up as in part (a) does not simplify the calculation without a calculator and as such students need to consider an alternative, namely the subtraction $(15.3 \times 10) - (15.3 \times 2)$.

TWM characteristics: Specialising   Convincing   Characterising   Critiquing   Improving

This question if written in a standard way could have been presented as follows:

Q1 a *The rectangle below is split in two with the dimensions shown.*

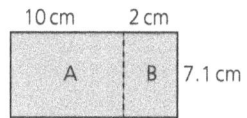

10 cm   2 cm

A   B   7.1 cm

  i *Calculate the area of rectangles A and B.*
 ii *What is the area of the rectangle with dimensions 12 cm × 7.1 cm?*
b *Work out the area of the rectangle below.*

15 cm

8.2 cm

Part (a) leads pupils to realise that the area of the whole rectangle can be calculated by splitting it up into two smaller rectangles, one of which has a length of 10 cm. Part (b) requires that pupils apply the same method to the new rectangle.

Support activity

Calculations with fractions

This activity provides additional practice of all required fraction calculations and should therefore be attempted after all exercises in Unit 15 have been completed.

1 Simplify these fractions.

a $\dfrac{80}{90}$

b $\dfrac{5}{10}$

c $\dfrac{25}{100}$

d $\dfrac{8}{12}$

e $\dfrac{15}{20}$

f $\dfrac{7}{21}$

2 Work out the following. Use the grid to help you.

a $\dfrac{9}{12}-\dfrac{2}{12}$

b $\dfrac{1}{4}+\dfrac{1}{2}$

c $\dfrac{1}{6}+\dfrac{2}{3}$

d $\dfrac{1}{3}+\dfrac{1}{2}$

e $\dfrac{2}{3}-\dfrac{1}{2}$

f $\dfrac{11}{12}-\dfrac{1}{2}$

3 Work out the following.

a $\dfrac{1}{5}$ of 20

b $\dfrac{1}{3}\times12$

c $60\times\dfrac{1}{10}$

d $\dfrac{1}{6}\times30$

e $\dfrac{1}{7}$ of 63

f $72\times\dfrac{1}{12}$

4 Write these improper fractions as mixed numbers.

a $\dfrac{8}{5}=1\dfrac{\square}{5}$

b $\dfrac{7}{2}=\square\dfrac{1}{2}$

c $\dfrac{11}{9}=\square\dfrac{2}{\square}$

d $\dfrac{13}{8}=\square\dfrac{\square}{8}$

e $\dfrac{11}{3}=3\dfrac{\square}{\square}$

f $\dfrac{17}{6}=\square\dfrac{5}{\square}$

5 Write these mixed numbers as improper fractions.

a $1\dfrac{3}{4}=\dfrac{\square}{4}$

b $1\dfrac{1}{2}=\dfrac{3}{\square}$

c $2\dfrac{3}{5}=\dfrac{\square}{5}$

d $2\dfrac{4}{7}=\dfrac{18}{\square}$

e $3\dfrac{1}{9}=\dfrac{\square}{9}$

f $3\dfrac{1}{8}=\dfrac{\square}{\square}$

6 Write each set of fractions in order of size.
Write the smallest fraction first.

a $\dfrac{5}{8}$ $\dfrac{7}{8}$ $\dfrac{3}{4}$

b $\dfrac{7}{10}$ $\dfrac{1}{2}$ $\dfrac{3}{5}$

c $\dfrac{3}{5}$ $\dfrac{11}{15}$ $\dfrac{2}{3}$

d $\dfrac{11}{12}$ $\dfrac{5}{6}$ $\dfrac{3}{4}$ $\dfrac{2}{3}$

7 Here are Sam's end of year exam results.
In which subject did he do best?
In which subject did he do worst?
Give reasons for your answers.

Hint: Rewrite each fraction as $\dfrac{\square}{100}$

Maths	$\dfrac{43}{50}$
History	$\dfrac{16}{20}$
Science	$\dfrac{21}{25}$
English	$\dfrac{7}{10}$

Answers

1 a $\dfrac{8}{9}$

b $\dfrac{1}{2}$

c $\dfrac{1}{4}$

d $\dfrac{2}{3}$

e $\dfrac{3}{4}$

f $\dfrac{1}{3}$

2 a $\dfrac{7}{12}$ **b** $\dfrac{3}{4}$ **c** $\dfrac{5}{6}$

   d $\dfrac{5}{6}$ **e** $\dfrac{1}{6}$ **f** $\dfrac{5}{12}$

3 a 4   **b** 4   **c** 6

   d 5   **e** 9   **f** 6

4 a $1\dfrac{3}{5}$ **b** $3\dfrac{1}{2}$ **c** $1\dfrac{2}{9}$

   d $1\dfrac{5}{8}$ **e** $3\dfrac{2}{3}$ **f** $2\dfrac{5}{6}$

5 a $\dfrac{7}{4}$ **b** $\dfrac{3}{2}$ **c** $\dfrac{13}{5}$

   d $\dfrac{18}{7}$ **e** $\dfrac{28}{9}$ **f** $\dfrac{25}{8}$

6 a $\dfrac{5}{8}$ $\dfrac{3}{4}$ $\dfrac{7}{8}$ **b** $\dfrac{1}{2}$ $\dfrac{3}{5}$ $\dfrac{7}{10}$

   c $\dfrac{3}{5}$ $\dfrac{2}{3}$ $\dfrac{11}{15}$ **d** $\dfrac{2}{3}$ $\dfrac{3}{4}$ $\dfrac{5}{6}$ $\dfrac{11}{12}$

7 Sam did best in maths and worst in English.

   Maths $\dfrac{86}{100}$,   History $\dfrac{80}{100}$,   Science $\dfrac{84}{100}$,   English $\dfrac{70}{100}$

Student's Book answers

Exercise 15.1 (page 118)

1

Fraction	Decimal	Percentage
$\frac{1}{2}$	**0.5**	**50%**
$\frac{1}{4}$	**0.25**	**25%**
$\frac{1}{5}$	0.2	**20%**
$\frac{1}{10}$	**0.1**	10%
$\frac{3}{8}$	0.375	**37.5%**
$\frac{21}{50}$	**0.42**	**42%**
$\frac{7}{40}$	**0.175**	17.5%
$\frac{16}{25}$	0.64	**64%**
$\frac{33}{250}$	**0.132**	13.2%

2 a $\frac{23}{25}$ 0.92   92%

$\frac{8}{5}$ 1.6   160%

$\frac{37}{20}$ 1.85   185%

$\frac{7}{8}$ 0.875   87.5%

b i $\boxed{\frac{17}{20}}$ ii 0.85 85%

3 a $\frac{61}{40}$ b $\frac{121}{400}$

c $\frac{513}{250}$ d $\frac{201}{2000}$

4 a With the numbers that are remaining the numerator is smaller than the denominator, therefore the answer is less than 1. The numerator cannot be 10, as there is only one zero available.

b The question states it's a fraction in its simplest form. If both numbers were even then both numerator and denominator would be divisible by 2 and therefore, not in its simplest form.

c $\frac{16}{25} = 0.64$

Exercise 15.2 (page 120)

1 a $\frac{17}{30}$ b $\frac{47}{60}$ c $\frac{5}{14}$

d $\frac{3}{26}$ e $\frac{11}{48}$ f 0

2 $\frac{18}{35}$

3 $\frac{9}{56}$

4 $\frac{3}{8} + \frac{1}{5}$

5 $\frac{11}{12} - \frac{2}{5}$

Exercise 15.3 (page 122)

1 a 3   b 4   c 3

2

3 a Whole number doubling each time, numerator staying the same and denominator doubling each time.

b $31\frac{31}{32}$

4 a a because only one fraction can be added to $3\frac{5}{8}$ to make $5\frac{1}{4}$

b Students' answers may vary.

Exercise 15.4 (page 126)

1 a $\frac{12}{35}$ b $\frac{1}{20}$ c $\frac{1}{60}$

d $\frac{1}{10}$ e $\frac{1}{6}$

2 $\boxed{\frac{1}{3}}$ $\boxed{3}$ $\boxed{\frac{3}{2}}$ $\boxed{\frac{2}{3}}$ $\boxed{\frac{3}{5}}$ $\boxed{\frac{5}{3}}$ $\boxed{\frac{1}{5}}$ $\boxed{5}$

3 a $\frac{1}{12}$ b $\frac{2}{27}$ c $\frac{4}{5}$

   d $\frac{5}{6}$ e $1\frac{1}{9}$

4 a True   b True   c True
   d True   e True

5 a $\frac{2}{3} \times \frac{7}{8}$ students' explanations

   b Students' answers may vary

Exercise 15.5 (page 128)

1 a Students' explanations
   b E.g. $15.3 \times (10 - 2) = 153 - 30.6 = 122.4$
      Students' explanations

2 Athlete B as $\frac{9}{16}$ is more than half, whilst $\frac{11}{24}$ is less than half.

3 a 45   b 164.7   c 102
   d 39   e 332.2   f $\frac{21}{32}$

4 Students' answers may vary, but are likely to be $\frac{8}{18}$ and $\frac{9}{18}$

5 a Q   b R

6 $\frac{15}{8}$ $\frac{11}{5}$ $\frac{9}{4}$ $2\frac{2}{5}$ $3\frac{1}{3}$

7 No, because $18\frac{2}{7}$ is equivalent to $18\frac{10}{35}$ and $18\frac{12}{35}$ is greater.

Workbook answers

Exercise 15.1 (page 47)

1
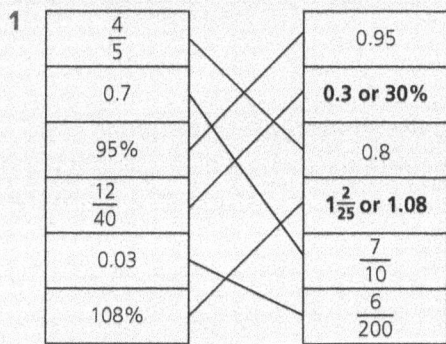

$\frac{4}{5}$	0.95
0.7	**0.3 or 30%**
95%	0.8
$\frac{12}{40}$	$1\frac{2}{25}$ or 1.08
0.03	$\frac{7}{10}$
108%	$\frac{6}{200}$

2 $\frac{2}{5}$ = 0.4 = 40%

   0.75 = $\frac{3}{4}$ = 75%

   65% = $\frac{13}{20}$ = 0.65

   $\frac{17}{20}$ = 0.85 = 85%

   1.02 = 102% = $\frac{51}{50}$

   $\frac{1}{3}$ = 0.3 = 33.3%

   30%, $\frac{1}{3}$, $\frac{2}{5}$, 65%, 0.75, $\frac{17}{20}$, 1.02

Exercise 15.2 (page 48)

1 $\frac{3}{10}$

2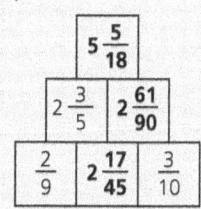

$\boxed{\frac{1}{6}}$ + $\boxed{\frac{3}{4}}$ = $\boxed{\frac{22}{24}}$

3 a $\frac{31}{60}$ and $\frac{1}{60}$ b $\frac{1}{60}$

Exercise 15.3 (page 49)

1 a

	$2\frac{41}{60}$	
$1\frac{11}{20}$		$1\frac{2}{15}$
$\frac{3}{4}$	$\frac{4}{5}$	$\frac{1}{3}$

   b

	$5\frac{5}{18}$	
$2\frac{3}{5}$		$2\frac{61}{90}$
$\frac{2}{9}$	$2\frac{17}{45}$	$\frac{3}{10}$

2 a Add $\frac{1}{4}$ b $1\frac{9}{10}$

Exercise 15.4 (page 49)

1 a $\frac{3}{8}$ cm²   b $2\frac{9}{56}$ cm²

Exercise 15.5 (page 50)

1 a $\frac{1}{7}$, $\frac{3}{14}$, $\frac{1}{2}$, $\frac{9}{10}$

   b $\frac{1}{3}$, $\frac{4}{9}$, $\frac{7}{10}$, $1\frac{2}{5}$

2 a 21   b 81
   c 84   d 54.6
   e 262.5

Probability and outcomes

Prior knowledge

Students will understand what mutually exclusive events are and will have had prior experience of conducting simple probability experiments.

Objectives overview

Learning objective	Objective code	*Student's Book* pages	*Workbook* pages	*Teacher's Guide* pages	Online resources
Identify all the possible mutually exclusive outcomes of a single event, and recognise when they are equally likely to happen.	7Sp.03	130–134	51–52	78–80	Flashcards Unit 16
Understand how to find the theoretical probabilities of equally likely outcomes.	7Sp.04	130–134	51–52	78–80	Knowledge test Unit 16 Worksheet: Two tone discs

Background information

In this unit, students build upon the foundations of their learning about probability, begun in Unit 10. They begin by defining mutually exclusive outcomes and use contexts such as dice, spinners and games to reinforce their understanding. Theoretical probability is revisited, and students will use the given formula to solve problems from a variety of contexts.

Terminology

Outcomes which cannot happen at the same time are called mutually exclusive.

Calculating the theoretical probability of equally likely outcomes can be written as a formula:

$$\text{Theoretical probability} = \frac{\text{Number of successful outcomes}}{\text{Total number of equally likely outcomes}}$$

Lesson ideas

The unit begins with revision of the term 'mutually exclusive outcomes' with examples. There is a reminder of the term 'theoretical outcomes' from Unit 10 with examples and exercises. This is a short and straightforward unit.

Starter activity

Plummet

Play a game of 'Plummet!' with a friend. You will need a 6-sided dice.

Player 1 rolls the dice.

Use the table to find out what the score means.

Score on dice	This means you...
1	Lose everything you have scored this turn. End your turn!
2	Add 0.2
3	Add 0.3
4	Add 0.4
5	Add 0.5
6	Subtract 0.6 if your total is 0.6 or more.

After each roll of the dice **player 1** needs to decide either to:
- BANK their total and end their turn
- or to THROW again.

When you BANK your score is safe and cannot be lost by throwing a '**1**' next turn.

The winner is the first person to get to 10.

Student's Book answers

Exercise 16.1 (page 131)

1 a 1, 2, 3, 4, 5, 6   b $\frac{1}{6}$

2 a Spinners A and B show equally likely outcomes because each colour is as likely as any of the other colours.
   b Spinner A, the probability of each colour is $\frac{1}{4}$
      Spinner B, the probability of each colour is $\frac{1}{2}$

3 a Black, Red, Blue
   b Unlikely to be as the colour the archer hits will be dependent on how good they are.
   c Yes, because the archer cannot hit two colours with one arrow.

   d He was as likely to hit the black as he was to hit red as both probabilities were 0.4.

4 a $1 - 0.22 - 0.44 - 0.30 = 0.04$
   b 0.66
   c No, because it is possible for a student to play both sports as shown in the overlap of the two circles.

5 a $\frac{14}{28} = \frac{1}{2}$
   b Girls + Long Jump, Girls + High Jump, Girls + Javelin, Boys + Long Jump, Boys + High Jump and Boys + Javelin.
   c No, because there are different numbers of students taking part for each outcome.
   d In this case Boys + High Jump are mutually exclusive as there are no boys doing it.

Exercise 16.2 (page 133)

1 Carla is wrong because, although there are three outcomes, they are not equally likely as it will depend on how good her team is compared with the opposition.

2 Manufacturer C's claim is the most reliable as it is based on the greatest number of results.

3 a The choosing of each fish may not be equally likely, e.g. the child may have gone into the shop to buy a red fish. This would therefore make the probability of choosing a red fish a lot greater.

  b For a fish to be randomly chosen.

4 a Not able to tell yet

  b Three tests is not enough to judge whether it is biased or not. The experiment needs to be repeated a lot of times.

5 The drawing pin needs to be dropped a lot of times (at least 100) and the results recorded. The probability that the pin lands point up can be calculated from the results as follows:

Probability of pin landing point up =
$$\frac{\text{Number of times pin landed point up}}{\text{Number of times the pin was dropped}}$$

6 a Round 6, because there are four cards left and two of them (7 and 9) are bigger than 5

  b Round 4, because there are six cards left and four of them (6, 7, 9 and 10) are bigger than 5 and $\frac{4}{6} = \frac{2}{3}$

Workbook answers

Exercises 16.1–16.2 (page 51)

1 a 2, 3, 5, 7 b $\frac{1}{4}$

2 a $\frac{4}{20}$ b $\frac{3}{20}$

3 a 0.11 b 0.77

  c No, as a student can choose to study both languages, shown by the overlap of the two circles.

4 a Cannot tell. The dice needs to be rolled more times to know whether or not it is biased.

  b Yes. The number 2 has been rolled more times than you would expect.

5 a Spin the spinner 60 times. If it was fair it would land on 1 roughly 20 times, 2 roughly 10 times and 3 roughly 30 times.

  b i $\frac{2}{6}$ ii $\frac{1}{6}$ iii $\frac{3}{6}$

17 Angle properties

Prior knowledge

Students have had experience working with angles and triangles in previous learning. They will be able to describe angles using the terms acute, obtuse, right or reflex and they will know that the sum of angles in a triangle is 180 degrees.

Objectives overview

Learning objective	Objective code	Student's Book pages	Workbook pages	Teacher's Guide pages	Online resources
Draw parallel and perpendicular lines, and quadrilaterals.	7Gg.14	135–152	53–56	81–86	Flashcards Unit 17
Derive the property that the sum of the angles of a quadrilateral is 360°, and use this to calculate missing angles.	7Gg.11	135–152	53–56	81–86	Knowledge test Unit 17 Worksheet: Clocks
Know that the sum of the angles around a point is 360°, and use this to calculate missing angles.	7Gg.12	135–152	53–56	81–86	
Recognise the properties of angles on: parallel lines and transversals; perpendicular lines; intersecting lines.	7Gg.13	135–152	53–56	81–86	

Background information

This unit begins with revision of measuring and drawing angles with examples. The unit shows, in detail, several constructions followed by an exercise.

Students will learn the sum of the angles of a quadrilateral using their knowledge of triangles. Angles formed between parallel lines are demonstrated though practical work to show vertically opposite, alternate and corresponding angles. The last question in this unit extends student understanding of tessellation, previously visited in Unit 13.

Terminology

Students should be familiar with the common 2D shapes studied in Unit 2, as their names and properties will aid their understanding of angles in these shapes.

Students know and use the term 'tessellate' and that shapes tesselate if they fit together without leaving any gaps.

Lesson ideas

Students are shown how to use a protractor properly and they will need opportunities to practise this skill. Students might work in pairs drawing and measuring acute, obtuse angles before working on Exercise 17.1.

Starter activity

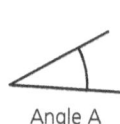

Optical illusions

1 **Without measuring:** Which of these angles is biggest?

a

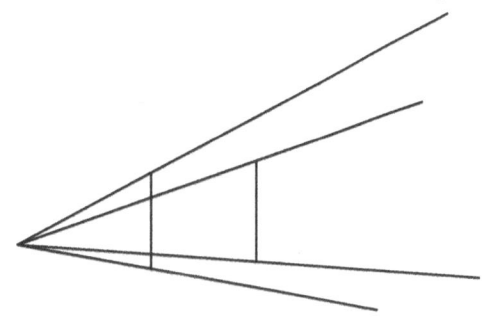

Angle A   Angle B

b

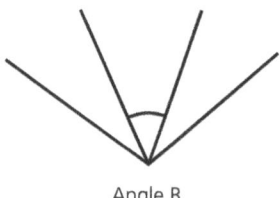

Angle A   Angle B

2 a Which of these vertical lines is longest?

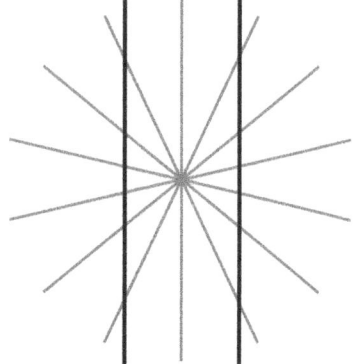

b Which of these horizontal lines is longest?

Line A   Line B

3 Are these vertical lines parallel?

a

b Are these horizontal lines parallel?

4 Look online for more examples of optical illusions.
Which one is your favourite?

Answers

1 a & b; both angles are the same size.
2 a & b; lines are the same length.
3 a & b; lines are parallel.

TWM activity notes

Exercise 17.4 is explained here as an exemplar of Thinking and Working Mathematically (TWM), detailed in the Introduction to the *Teacher's Guide*, page xi. Context: Students have covered the sum of angles around a point equals 360° and have been introduced to the properties of tessellating shapes.

Q4   *Shapes that tessellate fit together without leaving any gaps.*
For example, congruent squares and parallelograms tessellate with themselves as shown:

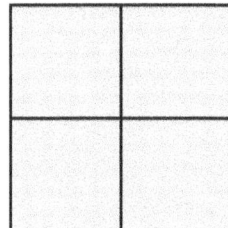 

Does the kite shown below tessellate with other congruent kites? Justify your answer.

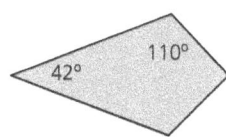

The question builds on the two simple examples of squares and parallelograms given, in that in order for the kites to tessellate, rotation of the shape is needed. The proof needed also draws on their knowledge of the sum of the four angles of quadrilateral, the angle properties of a kite in particular and of course the properties of angles around a point.

TWM characteristics: Specialising   Convincing   Characterising

This question if written in a standard way could have been presented as follows:

Q4   *Shapes that tessellate fit together without leaving any gaps.*
For example, congruent squares and parallelograms tessellate with themselves as shown:

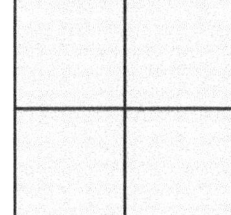 

The kite below is arranged with other congruent kites as shown.

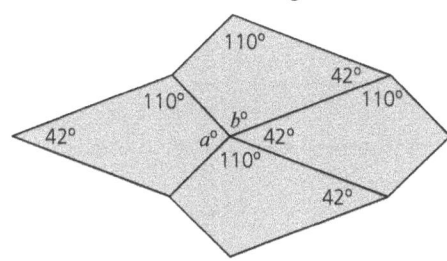

a Calculate angles a and b.
b Do the four angles 110°, 42°, a° and b° fit around a point exactly?

Here, students are given the arranged kites and prompted to find the missing angles a and b. As such, the task becomes a simple case of using the angle properties around a point adding up to 360°.

Extension activity

⭐ The clock's ticking!

1 What time does the clock show?

What is the angle between the hands?

2 Write down the angle between the hands at these times.

a

b

c

3 a Sam says the angle between the hands at 9.15 is 180°.
   Is Sam correct? Explain your answer.
  b Write down a time when the angle between the hands is 180°.
  c How many times is the angle between the hands 180° during a 12-hour period?
4 a Write down two times when the angle between the hands is 90°.
  b How many times is the angle between the hands 90° during a 12-hour period?

Answers

1 4 o'clock; 120°
2 a 30°   b 120°   c 60°
3 a No, as the hands aren't pointing exactly at 9 and 3
  b 6 o'clock
  c 11 times
4 a 9 o'clock and 3 o'clock
  b 22 times

Student's Book answers

Exercise 17.1 (page 136)

Student's measurements may differ by ±2°.
1 a 45°
   b 22°
   c 95°
   d 138°
   e 115°
   f 135°
2 a $a = 90°$; $b = 140°$; $c = 130°$
   b $d = 34°$; $e = 58°$; $f = 122°$; $g = 146°$
   c $h = 62°$; $i = 298°$
   d $j = 33°$; $k = 71°$; $l = 256°$
   e $m = 32°$; $n = 135°$; $o = 58°$; $p = 328°$
   f $q = 107°$; $r = 328°$; $s = 326°$; $t = 39°$
3 a–l Students' drawings of angles.

Exercise 17.2 (page 140)

1 & 2 Students' constructions of shapes.
3 a 45° + 30°   b 60° − 45°
   c 135° = 90° + 45°
      105° = 60° + 45°
      Other obtuse angles possible.

Exercise 17.3 (page 143)

1 $a = 360 − 70 − 50 − 135 = 105°$
2 a A rhombus
   b $b = 135°$ and $c = 45°$
      A rhombus has rotational symmetry of
      order 2, therefore diagonally opposite
      angles are equal.
3 a A parallelogram as it has two pairs of
      parallel sides.
   b $d = 180 − 125 = 55°$ (angles on a straight
      line = 180°)
      $e = 55°$ as it is equal to d (diagonally
      opposite)
      $f = g = 125°$ $\left(\dfrac{360 − 55 − 55}{2} = 125 \right)$
4 a It has one pair of equal opposite angles.
      Its diagonals intersect at right angles.
      It has two pairs of adjacent sides of the
      same length.
      It has one line of reflective symmetry.
      It has no rotational symmetry.
      Its internal angles add up to 360°.
   b $a = 20°$ horizontal line of symmetry
      $b = 100°$ horizontal line of symmetry
      $c = 120°$ angles of a quadrilateral sum
      to 360°

5 a Students' answers may vary but could
      include:
      A quadrilateral
      Two pairs of parallel sides
      Opposite diagonal angles equal
      Rotational symmetry of order 2
      Internal angles up to 360°
   b $e = 75°$ Opposite diagonal angles equal
      $d = f = \dfrac{360 − 75 − 75}{2} = 105°$ Opposite
      diagonal angles equal and sum of
      angles is 360°.
6 Angle BCD = angle CBA = 102.5° as
   triangle DCE is isosceles and angle DCE =
   $\dfrac{180 − 25}{2}$.
   Angle BAD = angle ADC = 77.5° as angles
   of trapezium add up to 360°.

Exercise 17.4 (page 146)

1 a $a = 110°$ b $b = 145°$
   c $c = 55°$ d $d = 95°$
   e $e = 100°$ f $f = 125°$
2 a $g = 106°$ b $h = 150°$
   c $i = 90°$ d $j = 60°$
3 Pieces A, B, D and E fit around a point as
   48 + 102 + 165 + 45 = 360°.
4 Yes because 110 + 110 + 42 + 98 = 360° as
   shown.

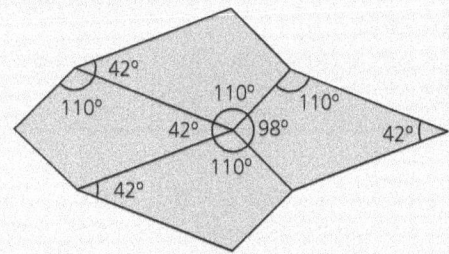

Exercise 17.5 (page 148)

1 a–c Students' drawings and measured
        angles.
   d Students' own observations leading to:
      vertically opposite angles are equal.

Exercise 17.6 (page 149)

1 a–c Students' drawings and measured
        angles.
   d Students' own observations leading to:
      corresponding angles are equal.

Exercise 17.7 (page 151)

1. a $a = 40°$; $b = 140°$
   b $c = 60°$; $d = 120°$
   c $e = 40°$; $f = 140°$
   d $g = 48°$; $h = 132°$
   e $j = 144°$; $k = 36°$
   f $l = 70°$; $m = 110°$
   g $n = 80°$; $o = 100°$; $p = 100°$; $q = 80°$.
   h $r = 43°$; $s = 137°$; $t = 137°$; $u = 43°$
2. a $v = 35°$; $w = 145°$; $x = 145°$; $y = 35°$;
      $z = 145°$
   b $a = 36°$
3. Angle ABD = 130°, therefore $x = 25°$

Angle DBC = 50° (angles on a straight line = 180°)
Angle BDC = 50° (if $x = 25°$, $2x = 50°$)
Angle DBC = angle BDC = 50°. Angle BCD = 80°, therefore triangle BCD is isosceles

4. a $x + y = 180°$ as angle IEF = x and angle AEF + IEF = $x + y = 180°$ as angles on a straight line = 180°.
   b Angles around any point are $2x + 2y = 360°$ therefore will tessellate.

Workbook answers

Exercises 17.1–17.2 (page 53)

1. a $p = 90°$; $q = 25°$; $r = 245°$
   b $m = 50°$; $n = 46°$
2. a, b Students' constructions drawn accurately – allow tolerance of ±2mm and ±1°
3. Students' constructions drawn accurately – allow tolerance of ±2mm and ±1°

Exercises 17.3–17.4 (page 54)

1. $a = 73°$ Opposite angles in a parallelogram are equal.
   $b = 107°$ Adjacent angles in a parallelogram sum to 180°.
   $c = 68°$ Angles in a quadrilateral sum to 360°.
2. $n = 12°$ $m = 73°$

Exercises 17.5–17.7 (page 55)

1. a $c = 54°$ Vertically opposite angles are equal
      $d = 126°$ Angles on a straight line sum to 180° or Angles around a point sum to 360°
   b $e = 114°$ Alternate angles
      $f = 66°$ Supplementary to angle f
2. $x = 147°$ Supplementary angles sum to 180°
   $y = 123°$ Angles in a triangle sum to 180° and angles on a straight line sum to 180°

18 Algebraic expressions and formulae

Prior knowledge

This unit builds on the work done in Unit 6: Algebra beginnings – Using letters for unknown numbers.

Objectives overview

Learning objective	Objective code	*Student's Book* pages	*Workbook* pages	*Teacher's Guide* pages	Online resources
Understand that a situation can be represented either in words or as an algebraic expression, and move between the two representations (linear with integer coefficients).	7Ae.04	153–157	57–59	87–90	Flashcards Unit 18 Knowledge test: Unit 18
Understand that a situation can be represented either in words or as a formula (single operation), and move between the two representations.	7Ae.05	153–157	57–59	87–90	

Background information

Building on their learning in Units 6 and 14, in this unit students learn to form expressions from a theoretical situation. Once formed, these expressions will be simplified. Students will learn how to derive a formula from written information, for example to calculate a currency exchange rate. Finally, students practise rearranging formulae.

 You could draw a link to using symbols and formulae to represent scientific ideas in Stage 7 Cambridge Lower Secondary Science, Models and representations, and also to collating and summarising ideas and information from text, Cambridge Lower Secondary English, Interpretation of texts.

Terminology

Teachers will need to explain and demonstrate the differences between an equation and a formula.

Lesson ideas

The unit revises work done in Unit 6 and Exercise 18.1 tests this. The unit builds on Unit 6 – deriving a formula for the perimeter of a rectangle – and extends this idea to derive other formulae. Exercise 18.2 builds on the work in the unit. The *Workbook* helps to reinforce these ideas.

Starter activity

 **Get to the point!**

1 Chen has *p* pencils.
Write down expressions for the number of pencils each of these people has.

	Expression
Ali has five pencils more than Chen	
James has four pencils fewer than Chen	
Amir has twice as many pencils as Chen	
Maya has half as many pencils as Chen	

2 Write down an expression for the number of pencils each of these people has.

Name	Number of pencils	Expression
Jack	Two lots of *p*	
Yousef	*p* divided by 2	
Aki	2 more than *p*	
Sam	The sum of 2 and *p*	
Jasmin	*p* more than *p*	
Chloe	The product of *p* and *p*	
Peter	2 less than *p*	
Isobel	Half *p*	
Lucy	The product of 2 and *p*	
Kyle	*p* more than 2	
Dara	The square of *p*	

3 Chen has six pencils.
How many pencils does each person have?
4 Which expressions are equivalent?

Answers

1

	Expression	Q3 Answer
Ali has five pencils more than Chen	$p + 5$	11
James has four pencils fewer than Chen	$p - 4$	2
Amir has twice as many pencils as Chen	$2p$	12
Maya has half as many pencils as Chen	$\dfrac{p}{2}$	3

2

Name	Number of pencils	Expression	Q3 Answer
Jack	Two lots of p	$2p$	12
Yousef	p divided by 2	$\dfrac{p}{2}$	3
Aki	2 more than p	$p + 2$	8
Sam	The sum of 2 and p	$p + 2$	8
Jasmin	p more than p	$p + p$	12
Chloe	The product of p and p	$p \times p$	36
Peter	2 less than p	$p - 2$	4
Isobel	Half p	$\dfrac{p}{2}$	3
Lucy	The product of 2 and p	$2 \times p$	12
Kyle	p more than 2	$2 + p$	8
Dara	The square of p	p^2	36

4 $2p = p + p = 2 \times p$ $p \times p = p^2$ $2 + p = p + 2$

TWM activity notes

Exercise 18.1 is explained here as an exemplar of Thinking and Working Mathematically (TWM), detailed in the Introduction to the *Teacher's Guide*, page xi. Students have encountered how to write algebraic expressions and how to substitute values into it.

Q9   *There are three jigsaw puzzle boxes. The first box has n pieces, the second has 3n pieces and the third has $2n + 40$ pieces.*
    a   *Is it possible to tell which box has the least number of pieces? Justify your answer.*
    b   *Is it possible to tell which box has the greatest number of pieces? Justify your answer.*

Both parts are seemingly asking the student the same sort of question. In reality, however, students need an understanding of the fact that the value of an expression changes when different numbers are substituted into it. Here students need to firstly appreciate that n can only ever be a positive number and that therefore n will always be smaller than both $3n$ and $2n + 40$. However, depending on the value of n the size of $3n$ in relation to that of $2n + 40$ will vary. Students may find out when is $3n = 2n + 40$.

TWM characteristics: Specialising   Conjecturing   Convincing   Characterising

This question if written in a standard way could have been presented as follows:

Q9   *There are three jigsaw puzzle boxes.The first box has n pieces, the second has 3n pieces and the third has $2n + 40$ pieces.*
    a   *If $n = 10$, arrange the boxes in order of how many pieces they have.*
    b   *If $n = 100$, arrange the boxes in order of how many pieces they have.*
    c   *Comment on any differences and/or similarities between your two answers.*

Through substitution of two different values for n, students will find that the box with n pieces has the least in both cases, whilst the other two boxes swap order. The question though does not ask them to explain *why* this happens.

Student's Book answers

Exercise 18.1 (page 154)

1 $t - 4$

2 a $2x + 6$ b $17

3 a $3m - 15$ b 51

4 a $4x - 200$ b ¥5000

5 a Perimeter of A = $8x + 14$
   Perimeter of B = $14x - 4$
   When $x = 2$, perimeter of A is bigger as
   $30 > 24$

   b When $x = 5$, perimeter of B is bigger as
   $66 > 54$

6 Student's own problem

7 Student's own problem

8 All three expressions are possible
   Student's own expression

9 a As n must be a positive integer, n must
   be smaller than both $3n$ and $2n + 40$.

   b No, it is not possible to tell which of $3n$
   and $2n + 40$ is bigger as it depends on
   the numerical value of n, e.g. if $n = 20$,
   then $2n + 40$ is bigger. If $n = 50$, then $3n$
   is bigger.

10 a Several answers possible. As x must
   be positive, due to the height of the
   second person being equal to x,
   Annie < Billy, Annie < David and
   Billy < David.

   b Nehreen could be the tallest or shortest
   depending on the value of x, e.g.
   If $x < 55$ then Nehreen is the shortest.
   If $x > 70$ then Nehreen is the tallest.

Exercise 18.2 (page 157)

1 a $m = 60h$

   b $m = 60 \times 4\frac{1}{2} = 270$

2 a $h = 24d$

   b $h = 24 \times 14 = 336$

3 a $s = 3600h$

   b $s = 3600 \times 2\frac{1}{2} = 9000$

   c $h = \dfrac{s}{3600}$ therefore $h = \dfrac{27\,000}{3600} = 7.5$

4 a $c = 100m$

   b $c = 100 \times 3.3 = 330$

   c $m = \dfrac{c}{100}$ therefore $m = \dfrac{4550}{100} = 45.5$

Workbook answers

Exercises 18.1–18.2 (page 57)

1 a $x - 7$ b $2x$ c $4x - 7$

2 a $f + 2$ b $3f + 6$ c $5f + 8$

3 a Rectangle B would have a   b B   c $x > 16.5$
   negative perimeter which
   is impossible.

4 a $C = 35h + 20$ b $125

5 a $3a + 3$ b $3a + 4$ c $22a + 12$

6 a $T = 0.5W + 0.75$ b 2 hours and 30 minutes.

19 Probability experiments

Prior knowledge

In previous learning, students have explored probability. They will be able to use the language of probability to describe events and outcomes and will have conducted simple probability experiments.

Objectives overview

Learning objective	Objective code	*Student's Book* pages	*Workbook* pages	*Teacher's Guide* pages	Online resources
Design and conduct chance experiments or simulations, using small and large numbers of trials. Analyse the frequency of outcomes to calculate experimental probabilities.	7Sp.05	158–161	60–61	91–93	Flashcards Unit 19 Knowledge test Unit 19

Background information

Unit 19 continues students' probability journey, having encountered mutually exclusive events (Unit 10) and theoretical probability (Unit 16). In this unit, students learn about experimental probability and the law of averages. They apply this thinking to make predictions of probability.

 Unit 19 is on experimental probability. The introduction is about volleyball. This has parallels with work on probability in Cambridge Lower Secondary Science Stage 7 Thinking and Working Scientifically – Models and representations and Scientific enquiry: analysis, evaluation and conclusions.

Terminology

Students encounter the term 'bias' in this unit which could be presented through the notion of a 'loaded dice'.

By allowing students extended opportunities to test a theory (for example if a coin was biased or not), there is opportunity to demonstrate the meaning of 'relative frequency' and therefore determine the 'experimental probability'.

Lesson ideas

Teachers might start by revising the idea of random events and asking for examples. The unit begins with the statement 'in the long run': discuss this.

Relative frequency and experimental probability are revised and Exercise 19.1 is an experiment for students to carry out. Students will compare their result with the theoretical probability. There is then a discussion part to the unit at the end.

Starter activity

 **A fair trial?**

1 How many 6's would you expect to get when you roll a six-sided dice 12 times? Make a prediction for each number and then roll a dice 12 times to see if you were right.

Score	Prediction	Tally	Frequency
1			
2			
3			
4			
5			
6			

2 Repeat your experiment, but this time roll the dice 60 times.

Score	Prediction	Tally	Frequency
1			
2			
3			
4			
5			
6			

Was your prediction correct?

3 Is your dice fair?

Can you prove the dice is fair?

4 Chloe rolls a dice three times and got three 6s.

Jack says that this proves the dice is biased towards getting a 6.

Is Jack right?

Answers

1 You might expect each score to turn up twice, but that is unlikely to happen in reality: one number may turn up three or four times and others not at all.

2 You might expect each score to turn up ten times, but that is unlikely to happen in reality.

3 The dice is likely to be fair even though the scores do not perfectly match the predicted frequency. To show the dice is fair you need to roll the dice many more times (say 1000) and you would expect each number to come up roughly the same number of times.

4 Jack is wrong, Chloe's dice could be fair and just have landed on 6 three times in a row by chance. If Chloe rolled it 100 times and got 100 6's then she could be fairly sure it was biased as the chance of that happening is vanishingly small.

Student's Book answers

Exercise 19.1 (page 159)

1 a, b, c Students' results from their experiment.

2 a, c, d Students' results from their experiment. Results and conclusions may vary.

b $\frac{1}{4}$

3 a, b, c Students' probabilities and conclusions. Theoretical probability is

HH = $\frac{1}{4}$, TT = $\frac{1}{4}$ and HT or TH (in any order) = $\frac{1}{2}$

d, e Students' probabilities and conclusions may vary.

4 b i $\frac{1}{6}$ ii two

a, c, d, e, f Students' answers will vary but should find that as the number of trials increases the relative frequency gets closer to the theoretical probability.

5 b ii Students' results are likely to vary each time as the results are down to chance.

a, bi, c, d Repeating the experiment more times should give students more confidence with their predictions.

Workbook answers

Exercise 19.1 (page 60)

1 a 50 times, probability of landing on heads is $\frac{1}{2}$

b Student to complete tally chart

c (number of times landed on heads) ÷ 100

d Fair – landed on heads close to 50 times
Unfair – didn't land on heads close to 50 times.

2 a Student to complete tally chart

b (number of times landed on 6) ÷ 30

c (number of times landed on 1, 3 or 5) ÷ 30

d Student compares their results to the theoretical probability $\left(\text{b. } \frac{1}{6}, \text{c. } \frac{1}{2} \right)$

e Roll the dice more than 30 times.

Introduction to equations and inequalities

Prior knowledge

This unit builds on the previous algebra Units 6, 14 and 18.

Objectives overview

Learning objective	Objective code	Student's Book pages	Workbook pages	Teacher's Guide pages	Online resources
Understand that a situation can be represented either in words or as an equation. Move between the two representations and solve the equation (integer coefficients, unknown on one side).	7Ae.06	162–167	62–64	94–96	Flashcards Unit 20
Understand that letters can represent an open interval (one term).	7Ae.07	162–167	62–64	94–96	Knowledge test Unit 20 End of Section 2 test

Background information

In this unit, students are shown equations as balanced scales. The aim is to reinforce the understanding that an equation represents two quantities which are equal to each other. Teachers can follow the detailed explanations and diagrams through with students to ensure understanding before students solve simple equations without visual representation. Students will learn how to form equations using written information, before then solving that equation.

Students will learn about inequalities and how these can be written and represented visually on a number line.

Terminology

Teachers might revise expressions, equations and formulae asking students to give examples of each.

Students should learn the inequality symbols, given in the *Student's Book* and presented here:

$>$ means 'is greater than'

$\geqslant$ means 'is greater than or equal to'

$<$ means 'is less than'

$\leqslant$ means 'is less than or equal to'

Lesson ideas

The inequality symbols could be displayed on a poster made by a student. The examples and exercise taken with those in the *Workbook* cover these ideas in detail. The unit can be taken as a detailed lesson plan.

Starter activity

 ## Ink splats

Aki's pen has leaked all over his maths homework.

Find the missing number under the ink splats.

1.	12 – ⬛ = 8	7.	9 – 2 × ⬛ = 1
2.	⬛ + 7 = 22	8.	10 – ⬛ = 13
3.	9 – ⬛ = 4	9.	⬛ ÷ 6 = 3
4.	⬛ × 4 = 12	10.	15 – 2 × ⬛ = 7
5.	15 ÷ ⬛ = 5	11.	⬛ – 6 × 3 = 7
6.	2 × ⬛ + 3 = 11	12.	12 – 2 × ⬛ = 22

Answers

1 $12 - 4 = 8$
2 $15 + 7 = 22$
3 $9 - 5 = 4$
4 $3 \times 4 = 12$
5 $15 \div 3 = 5$
6 $2 \times 4 + 3 = 11$

7 $9 - 2 \times 4 = 1$
8 $10 - (-3) = 13$
9 $18 \div 6 = 3$
10 $15 - 2 \times 4 = 7$
11 $18 - 6 \times 3 = 0$
12 $12 - 2 \times (-5) = 22$

Student's Book answers

Exercise 20.1 (page 164)

1 a $a = 2$
   b $b = 5$
   c $c = 7$
   d $d = 1$
   e $e = 8$
   f $f = 3$
   g $g = 8$
   h $h = 2$
   i $i = 2$
   j $j = 5$
   k $k = 5$
   l $l = 1$
   m $p = 1$
   n $q = 2$
   o $r = 17$
   p $s = 3$
   q $t = 19$
   r $u = 4$

2 a $v = 4$
   b $w = 7$
   c $x = 12$
   d $y = 10$
   e $z = 7$
   f $a = 2$
   g $b = 3$
   h $c = 6$
   i $d = 12$

   g $n = 10$ h $r = 7$
   i $q = 16$ j $k = 13$
   k $m = 12$ l $n = 7$
   m $r = 8$ n $q = 9$

Exercise 20.3 (page 165)

1 a $n + 10 = 24$ b $n - 7 = 7$
         $n = 14$ $n = 14$
   c $7n = 42$ d $n + 17 = 54$
         $n = 6$ $n = 37$
   e $n - 33 = 5$ f $11n = 132$
         $n = 38$ $n = 12$

2 $x = 64$
   Therefore, bottom floor holds 64 people,
   top floor holds 48 people.

3 7cm

Exercise 20.2 (page 164)

1 a $g = 5$ b $h = 1$
   c $i = 3$ d $j = 7$
   e $k = 5$ f $m = 7$

4 P has x sweets,   Q has $x + 5$ sweets
   R has $2x - 2$ sweets
   $4x + 3 = 83$
   $x = 20$
   Therefore, P has 20 sweets, Q has 25
   sweets and R has 38 sweets.

5 If perimeter is 24cm, then $x = 3$ making the
   shorter side 5cm long and therefore the
   area $5 \times 7 = 35$cm² not 30cm².

Exercise 20.4 (page 167)

1 a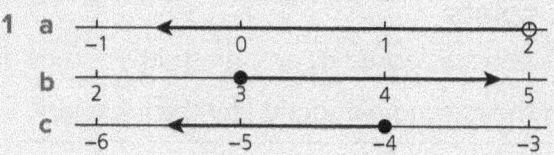
   b
   c
   d

2 a $x > 0$ b $x \leqslant 3$
3 a $n \leqslant 20000$ b $5x + 3 < 20$
   c $T \leqslant 25$ d $2x - 6 > 50$

Workbook answers

Exercises 20.1–20.2 (page 62)

1 a $m = 6$ b $n = 6$ c $t = 7$
   d $f = 6$ e $q = 4$ f $r = 25$

Exercise 20.3 (page 63)

1 a $x + 3 = 17$ $x = 14$
   b $5x = 65$ $x = 13$
   c $x - 18 = 4$ $x = 22$

2 50cm

Exercise 20.4 (page 64)

1 a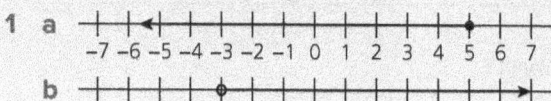
   b
   c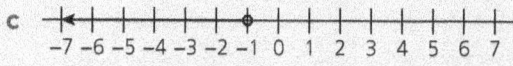
   d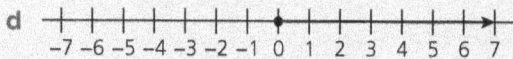

2 a $x \geqslant 1$ b $x < -1$
3 a $d \leqslant 3$ b $3x + 8 < 26$
   c $t \geqslant -6$

Section 2 – Review

1 Multiplying the unit digits of each number
   gives $6 \times 4 = 24$ therefore multiplying 36×24
   must give an answer ending in a 4 not a 6.

2 a i   7   ii 15
   b i   Not possible. For the median to be 8,
         the fifth card can have any number
         greater than or equal to 8.
      ii It can only be one of two numbers.
         14, because $14 - 3 = 11$ or 1 because
         $12 - 1 = 11$.

3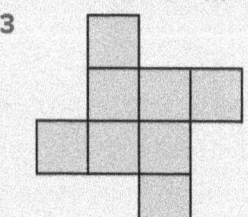

4 $6a + 14$
5 a $3\frac{7}{10}$ b $2\frac{3}{10}$ c $1\frac{1}{10}$
6 0.36
7 57° and students' explanations
8 a i   Sometimes true
      ii Only true if $x > 4$
   b i   Always true
      ii Area of A $= 12x + 18$, area of
         B $= 12x + 40$
         $12x + 18 < 12x + 40$ always
9 Yellow ≈ 16 Red ≈ 36 Green ≈ 29
10 a Students' explanations
   b 5

21 Sequences

 The Section 3 introduction talks about how geometry was used in Babylon, Egypt and Greece in ancient times. This could relate to Cambridge Lower Secondary English as students are encouraged to engage with the rich cultural roots of mathematics through the introduction.

Prior knowledge

Students should know their multiplication tables. Teachers can describe these as a sequence of numbers.

Objectives overview

Learning objective	Objective code	*Student's Book* pages	*Workbook* pages	*Teacher's Guide* pages	Online resources
Understand term-to-term rules and generate sequences from numerical and spatial patterns (linear and integers).	7As.01	172–178	47–50; 65–68;	97–100	Flashcards Unit 21 Knowledge test Unit 21 Worksheet: Grains of rice Worksheet: Matchstick patterns Worksheet: Tile patterns
Understand and describe *n*th term rules algebraically (in the form of $n \pm a$, $a \times n$ where a is a whole number).	7As.02	172–178	65–68	97–100	

Background information

In this unit, students begin by identifying and describing sequences and patterns they notice in numbers and shapes. They will extend sequences and learn to write a rule in words.

Concepts are further explored through term-to-term rules, with learning then extended to develop an understanding of the *n*th term and how it is calculated.

Terminology

Students will develop a clear understanding of the terms used to describe and use in the study of sequences: term, consecutive, extend, position, *n*th term.

Lesson ideas

The 'Let's talk' box suggests ways to begin the lesson. Multiplication tables allow for mental work as in 'What are the first four terms in the 6 times table?'

Sequences in diagrams gives an opportunity for students to make a poster. Exercise 21.1 tests this.

Term-to-term rule is explained in an example then Exercise 21.2 follows. Finally, the unit shows how to find a term-to-term rule and Exercise 21.3 follows.

The *Student's Book* and *Workbook* can be used as a detailed lesson plan if necessary.

Starter activity

Whatever next!

1 What comes next in these patterns? Explain how you know what comes next.

a ← ↑ → ↓ ← ↑ → ?

b Z Y X W V ?

c A C E G I K M ?

d M T W T F S ?

e M V E M J S U ?

f J F M A M J J ?

g Q W E R T ?

h | | 무 日 出 古 百 ?

2 This sequence is called a look-and-say sequence. Can you see why?

1, 11, 21, 1211, 111221, ?

a What is the next term?

b Will the digit '4' ever appear in this look-and-say sequence?

c A similar sequence starts with 2 instead of 1.
After the 2nd term how will each term end?

d Can you find a look-and-say sequence where every term is the same?

3 Think of some interesting patterns of your own and challenge a friend to find what comes next.

Answers

1 a ↓   (pattern is left, up, right, down, left, up, right down,...)

b U   (letters of the alphabet backwards)

c O   (alternate letters of the alphabet)

d S   (first letter of the days of the week)

e N   (first letter of planets in order of distance from Sun)

f A   (first letter of months of the year)

g Y   (letters on a keyboard, starting at top left)

h ⊞   (digital numbers and their reflection)

2 a This is called a 'look-and-say' sequence so you look at the previous term and say what you see to get the next term:

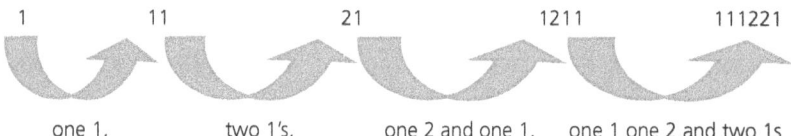

1   11   21   1211   111221

one 1,   two 1's,   one 2 and one 1,   one 1 one 2 and two 1s

So, the next term is: 312211 (three 1s, two 2s and one 1)

b There will never be a 4 or higher digit.

c 112

d 22, 22, 22, ...

Student's Book answers

Exercise 21.1 (page 173)

1 a

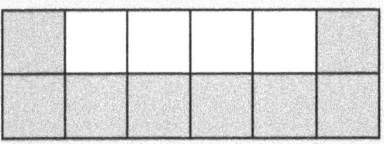

b

Number of white tiles	1	2	3	4	5
Number of red tiles	2	4	6	8	10

c The number of red tiles is double the number of white tiles.

d 200 red tiles

2 a

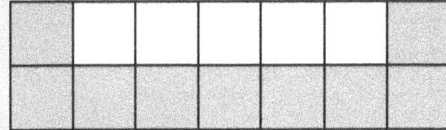

b

Number of white tiles	1	2	3	4	5
Number of green tiles	5	6	7	8	9

c The number of green tiles is 4 more than the number of white tiles.

d 104 green tiles

3 a

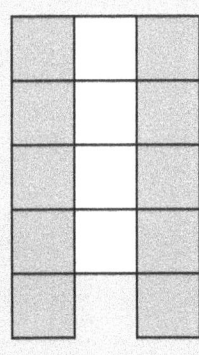
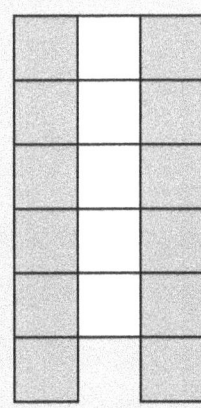

b

Number of white tiles	1	2	3	4	5
Number of orange tiles	4	6	8	10	12

c The number of orange tiles is double the number of white tiles, plus 2.

d 202 orange tiles

4 a

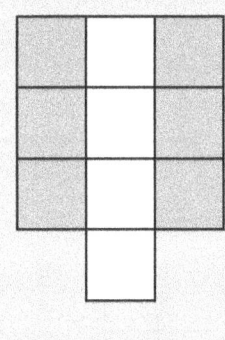
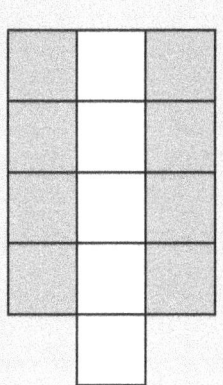

b

Number of white tiles	1	2	3	4	5
Number of blue tiles	0	2	4	6	8

c The number of blue tiles is double the number of white tiles, minus 2.

d 198 blue tiles

Exercise 21.2 (page 175)

1 a i +2 **ii** 12, 14 **iii** 20
  b i +2 **ii** 11, 13 **iii** 19
  c i +3 **ii** 19, 22 **iii** 31
  d i +4 **ii** 22, 26 **iii** 38
  e i +7 **ii** 36, 43 **iii** 64
  f i double each time
    ii 160, 320 **iii** 2560
  g i multiply by three
    ii 486, 1458 **iii** 39366
  h i −2 **ii** −1, −3 **iii** −9
  i i −4 **ii** 12, 8 **iii** −4
  j i −12 **ii** 96, 84 **iii** 36

Exercise 21.3 (page 177)

1 a i 11, 12 **ii** $n + 5$
  b i 14, 15 **ii** $n + 8$
  c i 12, 14 **ii** $2n$
  d i 48, 56 **ii** $8n$
  e i 0, 1 **ii** $n − 6$

f i 600, 700　ii 100*n*

g i −18, −21　ii −3*n*

2 a

Number of pink tiles	1	2	3	4	5
Number of blue tiles	3	6	9	12	15

b Number of blue tiles is three times the number of pink tiles.

c 3*n*　　d 195　　e 180

3 a 5, 10, 15, 20, 25　　b 5*n*

c 100

d Number of matchsticks must be a multiple of 5.

4 a (21, 60) as the *x*-coordinate is the triangle number + 1 and the *y*-coordinate is three times the triangle number.

b i Triangle 55 as the number is one less than the *x*-coordinate

ii 165

c 121. The triangle number is the *y*-coordinate ÷ 3, i.e. 120. The *x*-coordinate is the triangle number + 1.

Workbook answers

Exercise 21.1 (page 65)

1 a

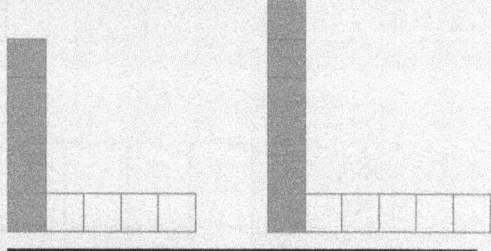

b

Number of white tiles	1	2	3	4	5
Number of grey tiles	2	3	4	5	6

c The number of grey tiles is always one more than the number of white tiles.

d 36

2 a

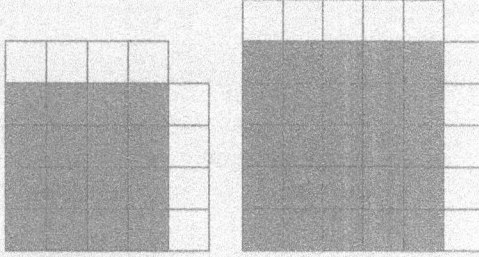

b

Number of white tiles	2	4	6	8	10
Number of grey tiles	1	4	9	16	25

c The number of grey tiles is the square of half the number of white tiles, and so is always a square number.

d 441

Exercise 21.2 (page 66)

1 a i +3　　ii 18, 21　　iii 30

b i +2　　ii 15, 17　　iii 23

c i +6　　ii 31, 37　　iii 55

d i −5　　ii −5, −10　　iii −25

2 a $55　　b 16 days

Exercise 21.3 (page 67)

1 a i *n* + 3　　ii 103

b i 4*n*　　ii 400

2 a 2*n* + 2

b Yes. The four corner squares will always be white, the centre white squares will always increase by 1.

3 a

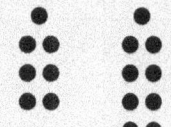

b 64　　　c 26th

4 a 4*a* + 7*b*

b 10*a* + 19*b*

Percentages of whole numbers

Prior knowledge

In previous learning, students have had experience of working with percentages. They will understand that percentages can be written as fractions with a denominator of 100 and will recognise percentages of shapes and whole numbers.

Objectives overview

Learning objective	Objective code	*Student's Book* pages	*Workbook* pages	*Teacher's Guide* pages	Online resources
Recognise percentages of shapes and whole numbers, including percentages less than 1 and greater than 100.	7Nf.05	179–184	69–71	101–103	Flashcards Unit 22 Knowledge test Unit 22

Background information

In this unit, students will learn how to find percentages of quantities. They begin by revising the link between percentages and their equivalent fractions. Students will find quantities as a percentage of another.

Finding percentages is a key life skill and therefore many of the questions found in the unit exercises are based on real-life scenarios.

Terminology

Revise the idea of a percentages as being 'out of 100'. The worked example uses a diagram with a detailed explanation as revision. Use the 'Let's talk' box with students. Exercise 22.1. follows on from this.

Lesson ideas

A quantity as a percentage of another. There is a worked example followed by Exercise 22.2.

A brief revision of what a Venn diagram is might be useful here. The *Student's Book* and the *Workbook* cover this topic to a good depth of learning.

Starter activity

Matching pairs

Match together equivalent fractions, decimals and percentages.

0.5	1	$\frac{3}{10}$	0.25	0.3
40%	75%	50%	0.2	0.6
$\frac{6}{10}$	$\frac{9}{20}$	$\frac{1}{5}$	0.75	20%
$\frac{3}{4}$	$\frac{1}{4}$	$\frac{1}{20}$	45%	$\frac{1}{2}$
0.05	0.45	30%	$\frac{2}{5}$	0.4
100%	25%	60%	$\frac{10}{10}$	5%

Answers

$0.5 = 50\% = \frac{1}{2}$

$1 = 100\% = \frac{10}{10}$

$\frac{3}{10} = 0.3 = 30\%$

$0.25 = \frac{1}{4} = 25\%$

$40\% = \frac{2}{5} = 0.4$

$75\% = 0.75 = \frac{3}{4}$

$0.2 = \frac{1}{5} = 20\%$

$0.6 = \frac{6}{10} = 60\%$

$\frac{9}{20} = 45\% = 0.45$

$\frac{1}{20} = 0.05 = 5\%$

Student's Book answers

Exercise 22.1 (page 180)

1 a 150   b 300   c 240   d 90
   e 200   f 40   g 90   h 770
2 612 are right-handed
3 84
4 3150
5 63
6 a

0.5%	$\frac{1}{200}$
100%	1
25%	$\frac{1}{4}$

125%	$\frac{5}{4}$
50%	$\frac{1}{2}$
150%	$\frac{3}{2}$

0.25%	$\frac{1}{400}$
$\frac{1}{3}\%$	$\frac{1}{300}$
30%	$\frac{3}{10}$

   b

0.5% × 600 = 3	125% × 600 = 750	0.25% × 600 = 1.5
100% × 600 = 600	50% × 600 = 300	$\frac{1}{3}$ × 600 = 2
25% × 600 = 150	150% × 600 = 900	30% × 600 = 180

7 a Students to show that eight squares are shaded each time.

b Students' explanations regarding $87.5\% = \dfrac{7}{8}$, leading to 49 shaded squares.

8 7 cats

9 54

10 a 10%   **b i** 10%

                                      ii $\dfrac{1}{10}$ is still shaded. Therefore 10% is shaded.

Exercise 22.2 (page 183)

1 a 50%	**b** 20%	**c** 25%
2 a 10%	**b** 60%	**c** 90%
3 a 125%	**b** 200%	**c** 500%
4 a 0.2%	**b** 0.1%	**c** 100%

5 Won 68%; Drew 8%; Lost 24%

6 87.5%

7 25%

8 a 6.25%   **b** 56.25%

9 80%

10 a 175%   **b** 75%

11 3.3% extra (1 d.p.)

12 20%

Workbook answers

Exercise 22.1 (page 69)

1 a 250   **b** 70   **c** 75

   d 450   **e** 240

2 341

3 $10.20

4 21

5

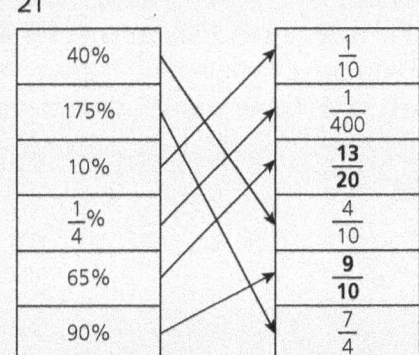

6 a

400				
80	80	80	80	80

b

160							
20	20	20	20	20	20	20	20

7 a 10 × 10% = 100% and 10 × 14 = 140

    5 × 20% = 100% and 5 × 14 = 70

b

180							
90				90			
22.5	22.5	22.5	22.5	22.5	22.5	22.5	22.5

Exercise 22.2 (page 71)

1 a 50%

   b 10%

   c 20%

   d 2.5%

2 Won = 50%; Drew = 37.5%; Lost = 12.5%

3 Black = 45%; Red = 30%; Blue = 25%

4 4%

Visualising three-dimensional shapes

Prior knowledge

Students have had previous experience working with 3D shapes. They will be able to recognise and draw nets for cuboids, prisms and pyramids.

Objectives overview

Learning objective	Objective code	Student's Book pages	Workbook pages	Teacher's Guide pages	Online resources
Visualise and represent front, side and top view of 3D shapes.	7Gg.08	185–191	72–74	104–108	Flashcards Unit 23 Knowledge test Unit 23

Background information

In this unit, students will practise visualising 3D shapes from different elevations. They will draw these elevations, starting with tangible objects they can see and use in the classroom such as cubes.

Students will consider questions about representations of 3D shapes such as, 'How many cubes would make this shape?'

Terminology

Students will be introduced to the terms plan and elevation. There are different elevation terms which must be learned and can be practised with objects in the classroom before progressing to theoretical drawn objects.

Lesson ideas

If possible, some architect drawings or house plans would be useful here. If teachers can get some Lego or similar blocks this topic becomes much easier to teach. Then the examples in the exercise can be seen in 3D. Students may bring in to class pyramid, cuboid and cylinder shapes and others if asked.

Starter activity

Flipped learning task – Ideal homes

Design your perfect house.

Draw a picture of your house in 3D.

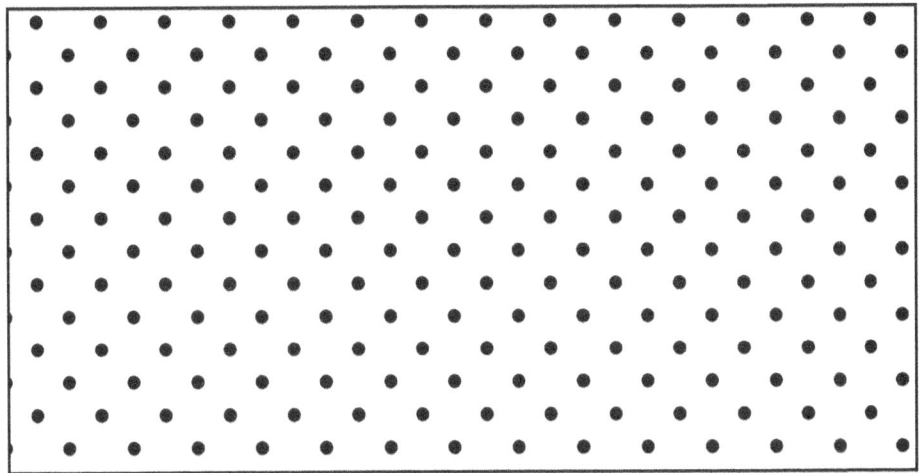

Now draw a picture of your house from

i above (a bird's eye-view – this is called a **plan**)

ii front-on (this called the **front elevation**)

iii side-on (this is called the **side elevation**)

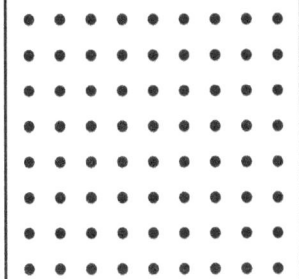

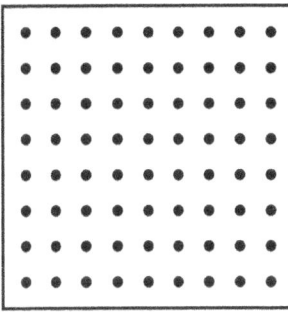

Student's Book answers

Exercise 23.1 (page 188)

1 a Students' explanations
   b

2 a C
   b

3 a x   y
   b i A or D
      ii   or

4 a 8
   b

5 a   b

Exercise 23.2 (page 190)

1 A circle
2 a   b

3 a   b
   c

4 a   b

5 a A regular hexagon
   b
   c Yes, because the sloping sides will be visible from above

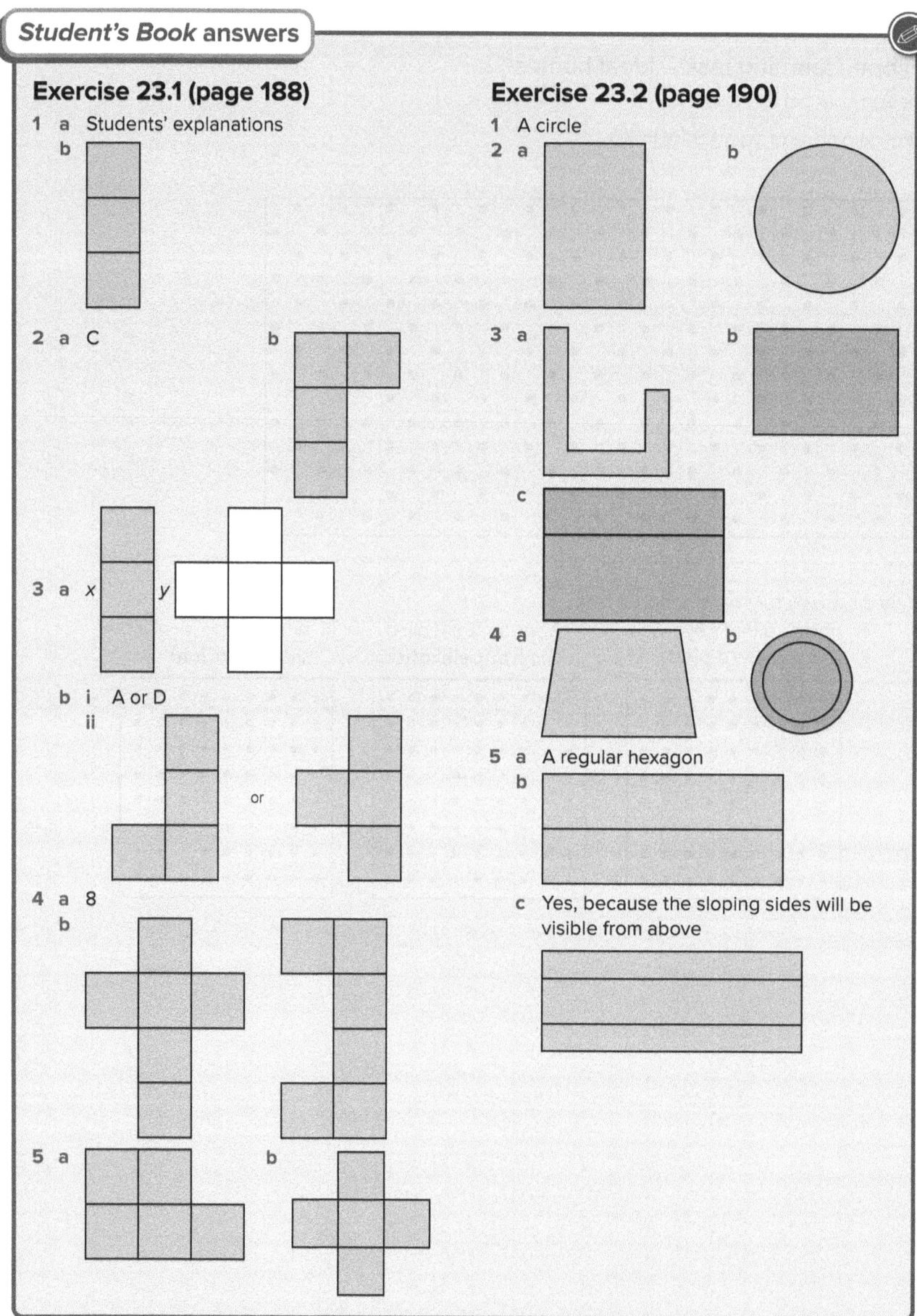

Workbook answers

Exercises 23.1–23.2 (page 72)

1

3D shape	Front	Side	Plan

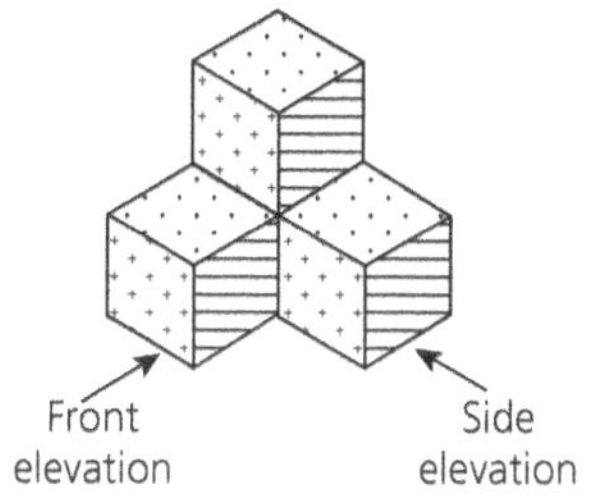

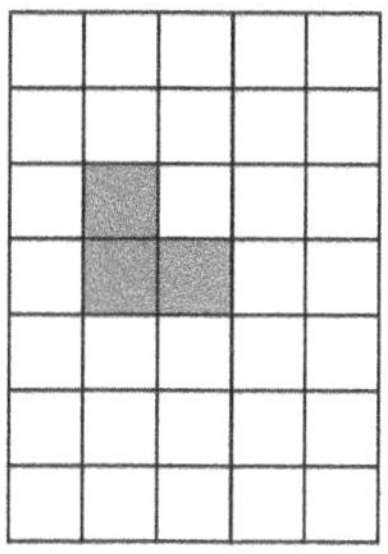

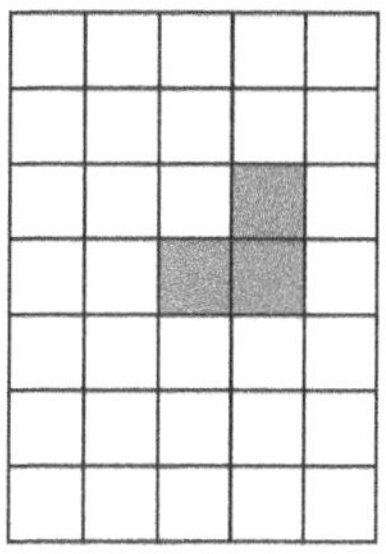

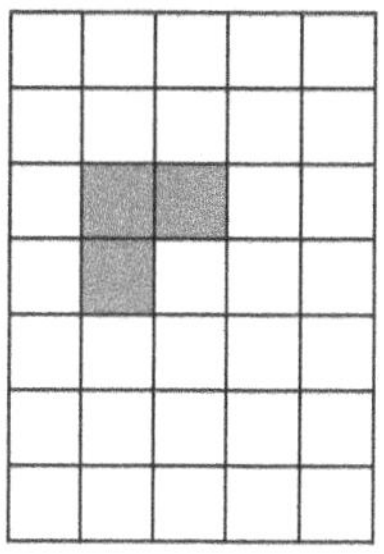

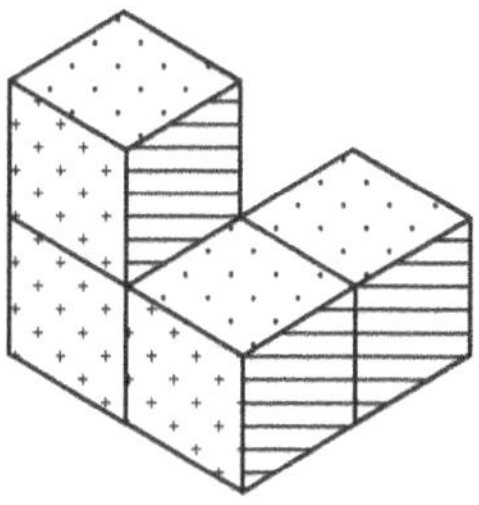

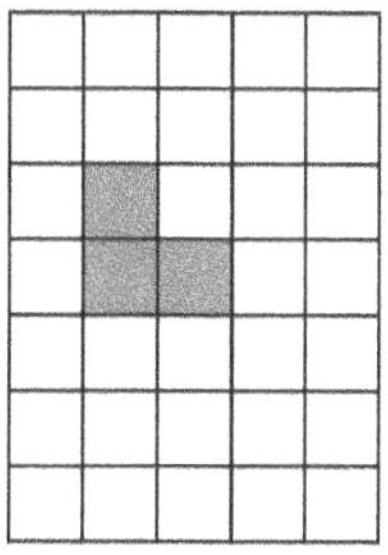

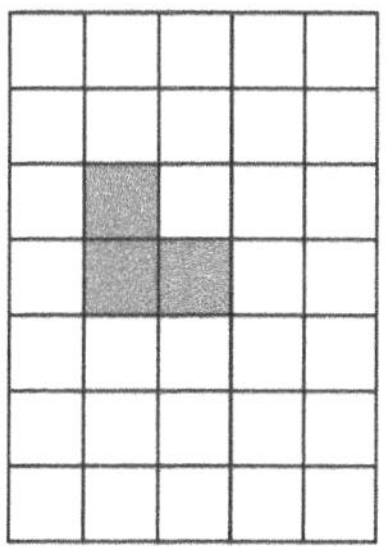

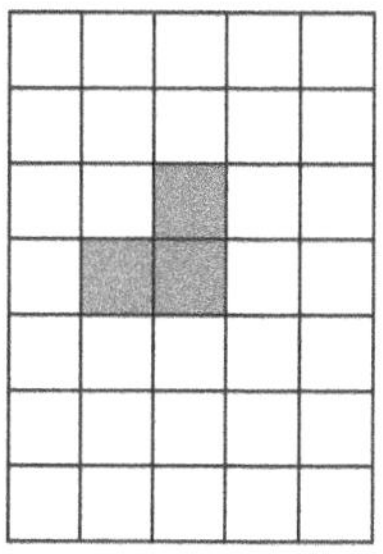

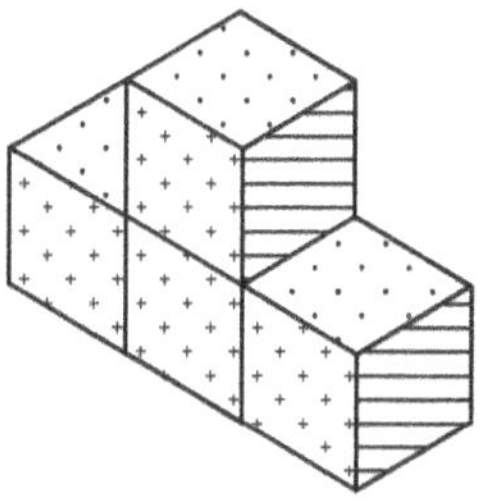

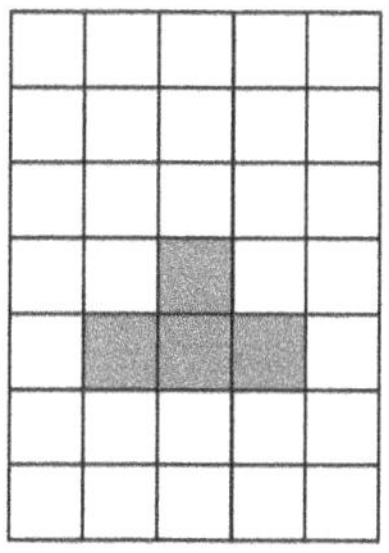

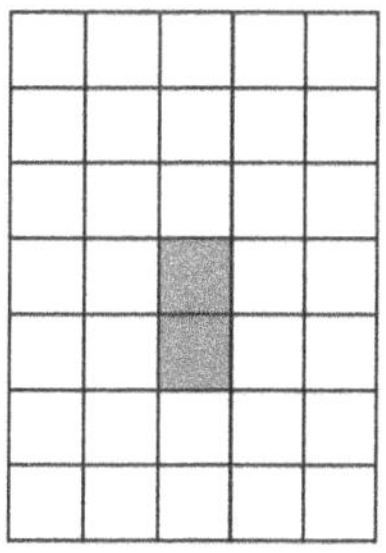

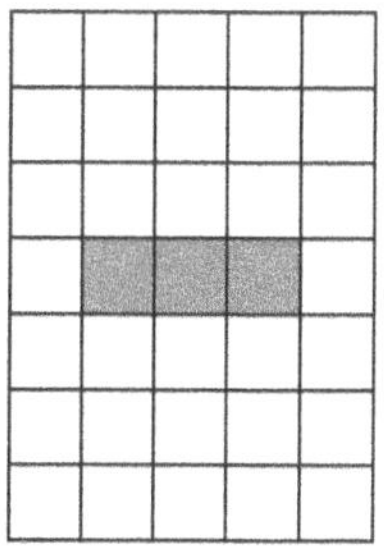

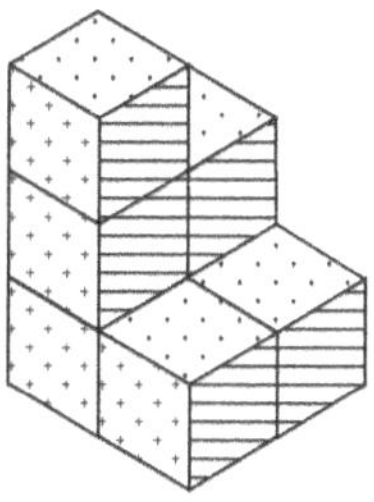

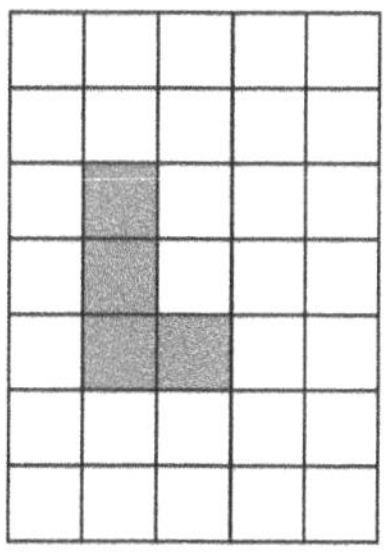

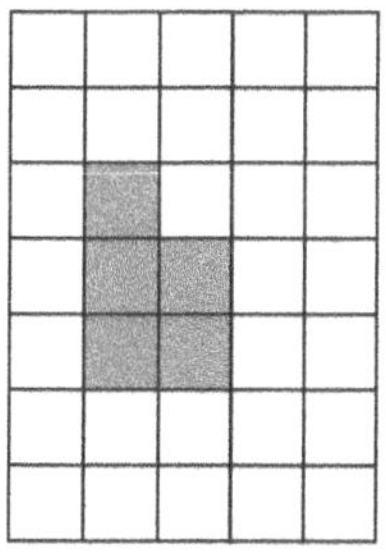

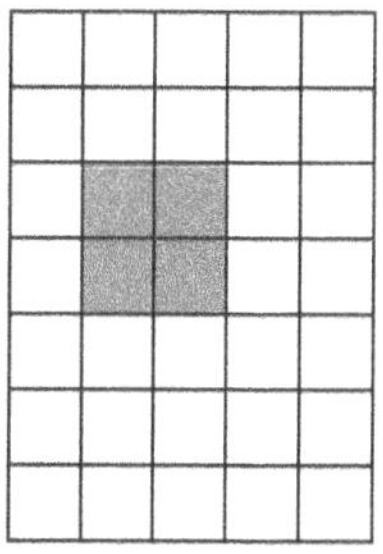

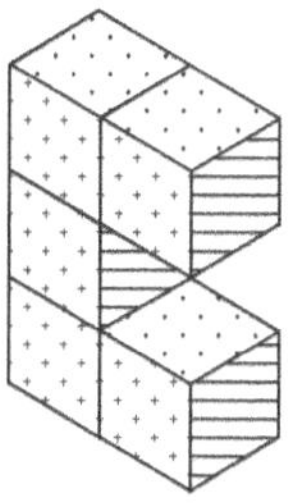

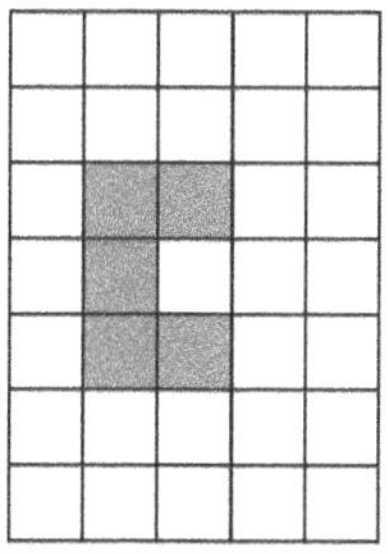

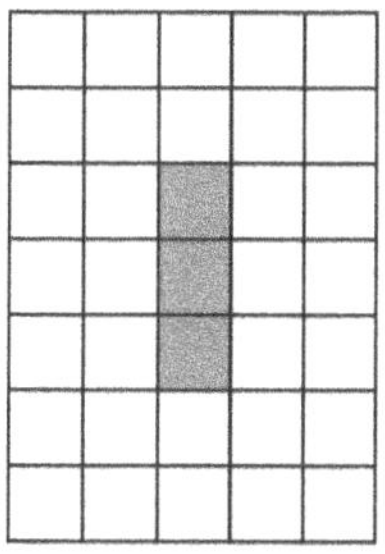

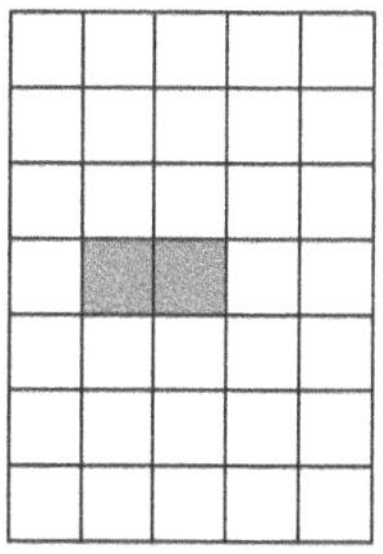

2

Front	Side	Plan	Mathematical name
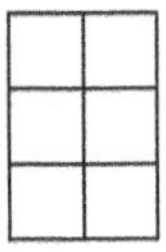	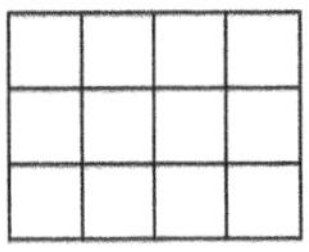	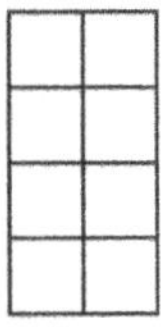	Cuboid
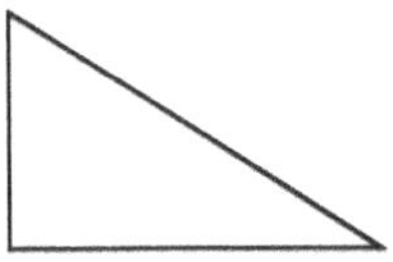			Triangular prism
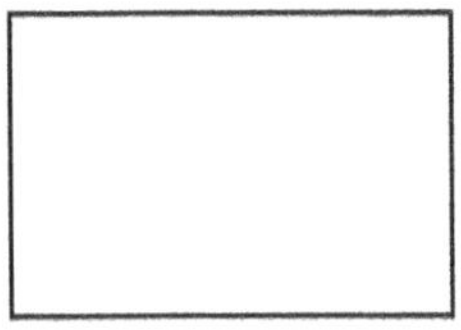	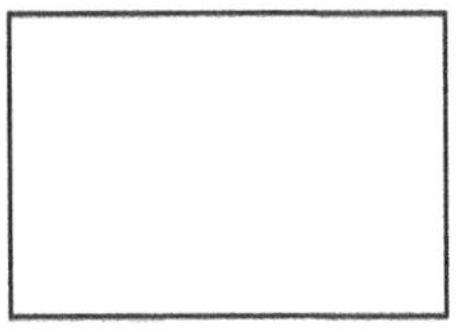	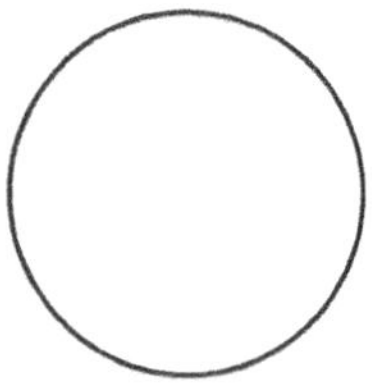	Cylinder

3 a

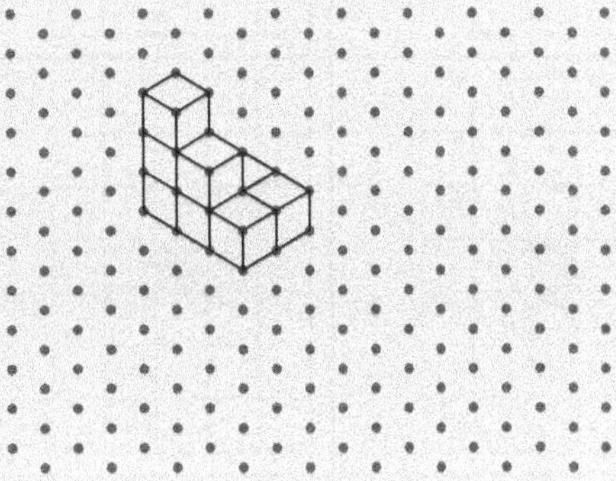

b 9

24 Introduction to functions

Prior knowledge

This is a new concept. It builds on the earlier algebra units in Sections 1 and 2 of the *Student's Book*.

Objectives overview

Learning objective	Objective code	*Student's Book* pages	*Workbook* pages	*Teacher's Guide* pages	Online resources
Understand that a function is a relationship where each input has a single output. Generate outputs from a given function and identify inputs from a given output by considering inverse operations (linear and integers).	7As.03	192–197	75–76	109–112	Flashcards Unit 24 Knowledge test Unit 24
Understand that a situation can be represented either in words or as a linear function in two variables (of the form $y = x + c$ or $y = mx$), and move between the two representations.	7As.04	192–197	75–76	109–112	

Background information

In this unit, students are introduced to the concept of functions. Study begins with function machines in which it is demonstrated that a number is input, the function carried out and produces an output. Students will carry out functions and use their problem-solving skills to identify the function when given the inputs and outputs. They will also explore inverse functions.

By the close of the unit, students will link their understanding of functions to forming equations, as studied in Unit 20.

Terminology

Explain the difference between 'a map' and 'mapping' in mathematics. The inverse of a function simply undoes the effect of the original. Make it very clear that an equation is an algebraic function. The exercises all show function machines.

Lesson ideas

Draw examples of function machines with simple inputs on the board. Students could play 'Guess my function?' by providing inputs and outputs for a partner. Students could be stretched further by providing the related inverse function.

Starter activity

Number machine

Professor Algebra has a number machine which can turn one number into another number.

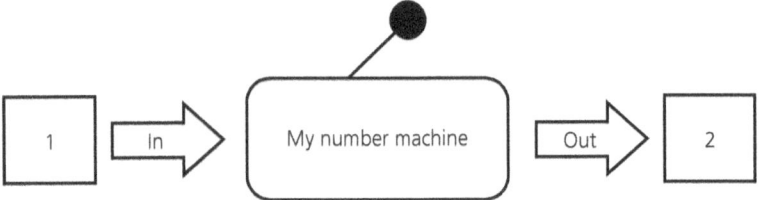

When Professor Algebra feeds a card with the number 1 into his machine, the number 2 comes out.

1 What rule could the machine be using?
2 Professor Algebra feeds some more cards into the machine.
   Here are his results.
   What rule is the machine using?

Number in	1	3	0	9
Number out	2	6	0	18

3 Professor Algebra re-sets his number machine and changes its rule.
   What rule is the machine using for each of the following sets of results?

   a

Number in	5	9	13
Number out	7	11	15

   b

Number in	10	20	30
Number out	2	4	6

   c

Number in	7	8	9
Number out	4	5	6

   d

Number in	20	40	60
Number out	15	30	45

4 Make up a rule for the number machine.
   Ask a partner to give you some numbers to feed into your machine.
   Tell your partner the results.
   Can they guess your rule? How many goes does it take?
   Is it possible to be sure of the rule after feeding in only one number?

Answers

1 'Add 1' or 'multiply by 2' or a more complicated rule like 'multiply by 2 and subtract 1'
2 Multiply by 2
3 a Add 2
  b Divide by 5
  c Subtract 3
  d Multiply by 0.75
4 Student's own work. It is not possible to be 100% sure of a rule after feeding in only one number; you need to feed in at least two numbers.

Students' Book answers

Exercise 24.1 (page 194)

1 a
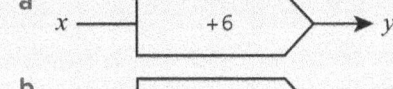
$x \longrightarrow \boxed{+6} \longrightarrow y$

  b
$q \longrightarrow \boxed{-5} \longrightarrow p$

2

Input	Output
0	5
1	6
2	7
3	8

3

Input	Output
2	6
4	12
6	18
8	24

4

Input	Output
2	1
4	2
6	3
8	4

5 a

In ◄─ Add 7 ─ Out

b

Input	Output
5	−2
23	16
15	8
119	112

6 a
In ◄─ Divide by 3 ─ Out

In ◄─ Subtract 8 ─ Out

b

Input	Output		Input	Output
−2	−6		−2	6
4	12		4	12
8	24		8	16
20	60		20	28

7 a 3   b 12

Exercise 24.2 (page 197)

1 $n = M - 56$

2 a
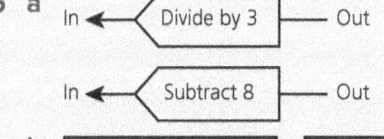
$h \longrightarrow \boxed{\times 15} \longrightarrow p$

  b $330

3 a
$R \longrightarrow \boxed{\times 17.4} \longrightarrow r$

  b 8700 Rupees   c 14.37 Reals

4 a
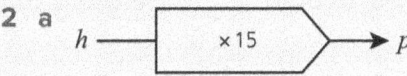
$n \longrightarrow \boxed{\times 0.8} \longrightarrow C$

  b Students' explanations. Forgetting to convert cents to dollars, cost can't be the input, it must be the output, +80 just adds 80 to the number of km.

  c $44

Workbook answers

Exercises 24.1–24.2 (page 75)

1 a

x —— [+7] —— w

b f —— [−3] —— d

c

h —— [×2] —— v

2 a

Input	Output
1	**2**
2	**4**
3	**6**
4	**8**

b

Input	Output
−2	**−6**
−1	**−3**
0	**0**
1	**3**
2	**6**

3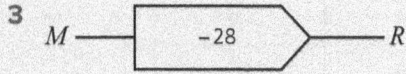

M —— [−28] —— R

4 a

In ←— [×2] —— Out

b

Input	Output
2	**1**
10	5
18	**9**
−6	−3
1	**0.5**

5 a

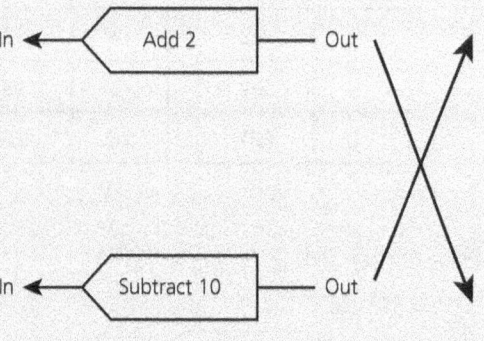

In ←— [Add 2] —— Out

In ←— [Subtract 10] —— Out

Input	Output
10	20
2.5	12.5
14	24
25	**35**
15	**25**

Input	Output
3.5	1.5
4	2
4.5	2.5
8	6
5.5	3.5

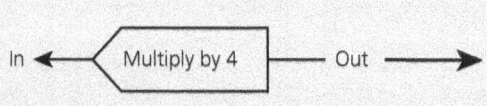

In ←— [Multiply by 4] —— Out —→

Input	Output
2	**0.5**
40	10
8	**2**
100	25
240	**60**

6 Input = x

$3x = x + 8$

$2x = 8$

$x = 4$

This proves there is only one possible solution and it is 4.

Coordinates and two-dimensional shapes

Prior knowledge

In previous learning, students have worked with 2D shapes on coordinate grids of up to four quadrants. They will be able to plot points on grids to draw lines and shapes. They will be able to translate shapes on a coordinate grid and identify corresponding points between the original shape and the translation.

Objectives overview

Learning objective	Learning objective	*Student's Book* pages	*Workbook* pages	*Teacher's Guide* pages	Online resources
Use knowledge of 2D shapes and coordinates to find the distance between two coordinates that have the same *x*- or *y*-coordinate (without the aid of a grid).	7Gp.02	198–204	77–80	113–119	Flashcards Unit 25

Knowledge test Unit 25

Worksheet: Coordinate patterns |
| Use knowledge of translation of 2D shapes to identify the corresponding points between the original and the translated image, without the use of a grid. | 7Gp.03 | 198–204 | 77–80 | 113–119 | |

Background information

In this unit, students begin by recapping the plotting of coordinates in all four quadrants. They plot missing coordinates to produce 2D shapes and find coordinates on diagonals and intersections. Students will consider coordinate changes to translated shapes.

Terminology

Language describing shape and movement features heavily in this topic. Students will need to have a good understanding of the language used to describe the properties of 2D shapes.

Lesson ideas

Exercise 25.1 is a revision exercise. This is followed by rhombus explanations regarding coordinates. Students are encouraged to discuss kites and arrowheads as well as rhombuses.

Translation is a sliding movement. Students need to understand this perhaps by cutting out and sliding shapes on a grid. Exercise 25.3 is difficult and some students may need help from the teacher. The unit is set out as a series of lesson plans in detail.

Starter activity

Hidden messages

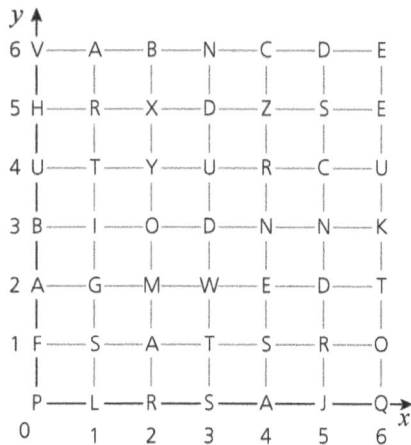

1 Use the grid to decode this question and its answer.

Question:

Start at (3, 2)	Go 3 left and 3 up	Go 2 right and 4 down	Go 4 right and 1 up		Go 3 left and 2 down
				,	

Start at (6, 1)	Go 2 left and 3 up	Go 4 left and 2 down	Go 3 right and 4 up	Go 4 down and 2 left	Go 3 right

Start at (4, 0)		Go 3 up		Go 1 right and 1 down	

Start at (3, 0)	Go 1 left and 3 up	Go 1 right and 1 up	Go 2 right and 1 down	Go 1 down	Go 2 left and 2 down

Start at (1, 0)		Go 3 up		Go 5 right		Go 2 up

Go to (0, 2)

Start at (0, 0)	Go 2 up	Go 2 right and 2 down	Go 3 right and 1 up	Go 1 right	Go 1 up	
						?

Answer:

Go to (2, 1)

Start at (5, 4)	Go 4 left and 2 up	Go 1 down	Go 3 right and 1 down	Go 2 left and 1 down	Go 4 right and 1 down	
						!

2 Make up your own hidden message and give it to a friend to decode.

Answers

1 Question:

Start at (3, 2)	Go 3 left and 3 up	Go 2 right and 4 down	Go 4 right and 1 up		Go 3 left and 2 down
W	h	a	t	'	s

Start at (6, 1)	Go 2 left and 3 up	Go 4 left and 2 down	Go 3 right and 4 up	Go 4 down and 2 left	Go 3 right
o	r	a	n	g	e

Start at (4, 0)	Go 3 up	Go 1 right and 1 down
a	n	d

Start at (3, 0)	Go 1 left and 3 up	Go 1 right and 1 up	Go 2 right and 1 down	Go 1 down	Go 2 left and 2 down
s	o	u	n	d	s

Start at (1, 0)	Go 3 up	Go 5 right	Go 2 up
l	i	k	e

Go to (0, 2)
a

Start at (0, 0)	Go 2 up	Go 2 right and 2 down	Go 3 right and 1 up	Go 1 right	Go 1 up	
p	a	r	r	o	t	?

Answer:

Go to (2, 1)
A

Start at (5, 4)	Go 4 left and 2 up	Go 1 down	Go 3 right and 1 down	Go 2 left and 1 down	Go 4 right and 1 down	
c	a	r	r	o	t	!

TWM activity notes

Exercise 25.3 is explained here as an exemplar of Thinking and Working Mathematically (TWM) as detailed in the Introduction to the *Teacher's Guide* page xi. Students will have covered coordinates in all four quadrants as well as the properties of quadrilaterals.

Q1 *The axes below show triangle ABC and its position to A'B'C' after a translation.*

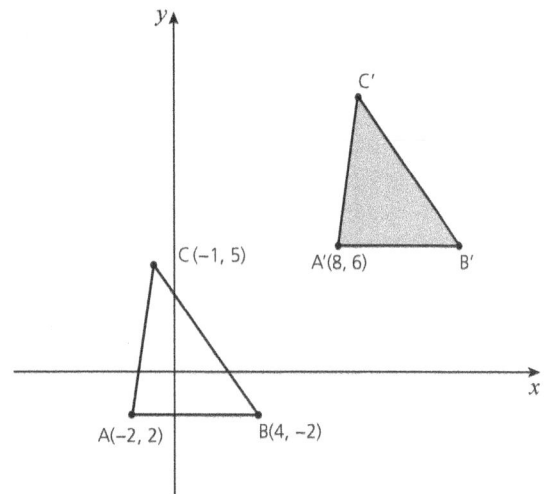

a Describe the translation that maps ABC on to A'B'C'.
b Deduce the coordinates of B'.
c Deduce the coordinates of C'.
d Explain why the area of triangle A'B'C' is 21 units².

Parts (a)–(c) are standard questions. However, part (d), by giving the solution, is placing the emphasis on the explanation in addition to drawing on other areas of mathematics. In order to confirm that the area of the triangle is 21 units², students need to realise that the base length is the difference between the x-coordinates of A' and B', whilst the height will be the difference between the y-coordinates of C' with either A' or B'.

TWM characteristics: Generalising   Convincing   Characterising

This question if written in a standard way could have been presented as follows:

Q1 *The axes below show triangle ABC and its position to A'B'C' after a translation.*

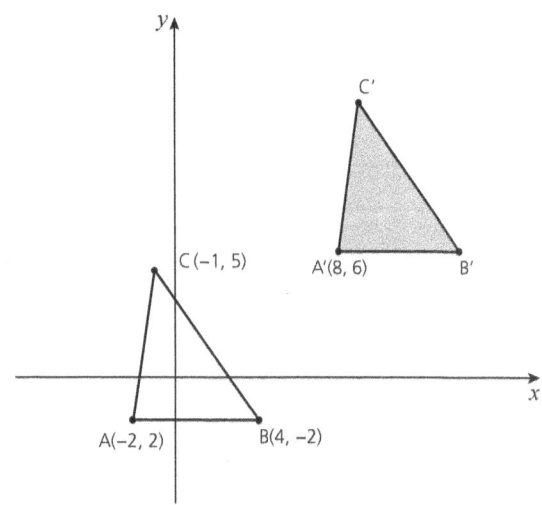

a Describe the translation that maps ABC on to A'B'C'.
b Deduce the coordinates of B'.
c Deduce the coordinates of C'.
d Calculate the length A'B'.
e Calculate the height of the triangle A'B'C'.
f Calculate the area of the triangle A'B'C'.

The end result is the same as for the TWM question, however here students are prompted to find the necessary dimensions in order to calculate the area.

Student's Book answers

Exercise 25.1 (page 199)

1

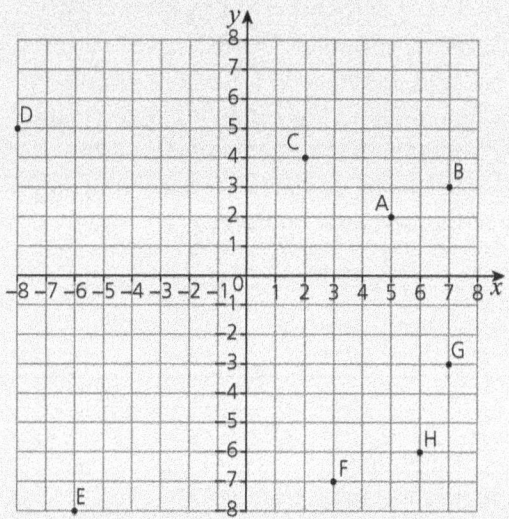

2 Rectangle

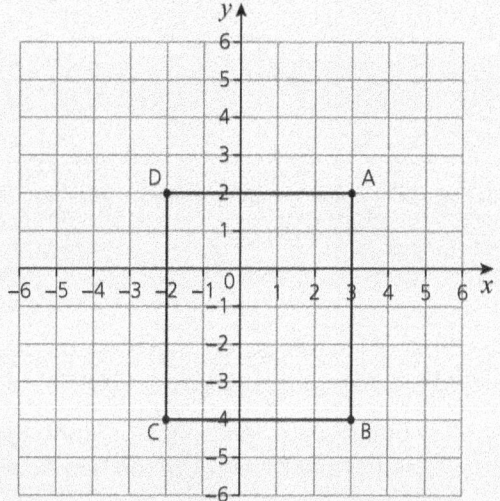

3 Isosceles triangle

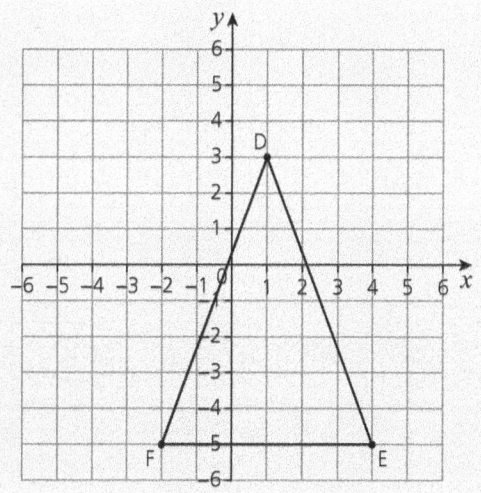

4 Parallelogram

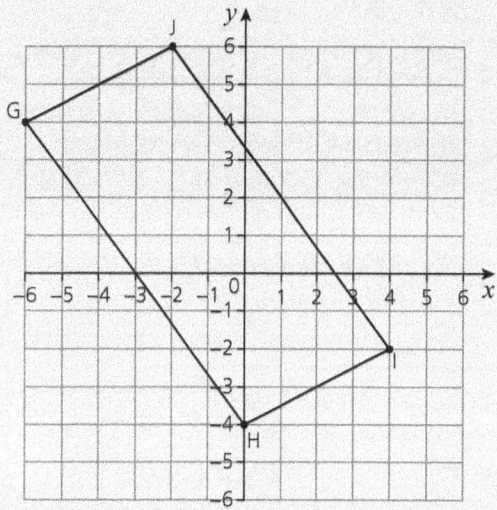

Exercise 25.2 (page 200)

1 a (4, 6)

b Has the same *x*-coordinate as A and the same *y*-coordinate as C.

2 a It is wrong because it must have the same *y*-coordinate as vertex Y, i.e. −3 not −5.

b Z (5, −3)

3 a i Student's plot

ii S(−6, −2)

iii Diagonals cross at (0, 1)

iv 72 units²

b i Student's plot

ii Parallelogram

iii 72 units²

iv It has the same area as the rectangle PQRS, i.e. the shape of the parallelogram does not affect its area.

c i Student's plot

ii J(0, 10)

iii 0

4 a R (0, 3); S (−3, 0)

b R (6, −3); S (3, −6)

5 a −3 as it must be in the same vertical line as vertex B.

b D (−3, −6)

6 N (2, −7); O (−2, −8)

Exercise 25.3 (page 203)

1 a 10 units to the right and 8 units vertically up

  b B (14, 6)

  c C (9, 13)

  d The base length is the difference between the x-coordinates of vertices A and B, i.e. 6 units. The height is difference between the y-coordinates of vertex C and either A or B, i.e. 7 units.

    Therefore area $= \frac{1}{2} \times 6 \times 7 = 21$ units².

2 a Q' is incorrect

  b The other vertices have been translated 4 units to the right and 4 units vertically up.

3 a E' (−3, 3.5); F' (−1, −4.5); G' (3, −3.5); H' (1, 4.5)

  b M is originally at (−5, −3.5) therefore to move to the origin it has translated 5 units to the right and 3.5 units vertically up. Each vertex is translated by the same amount.

Workbook answers

Exercises 25.1–25.2 (page 77)

1

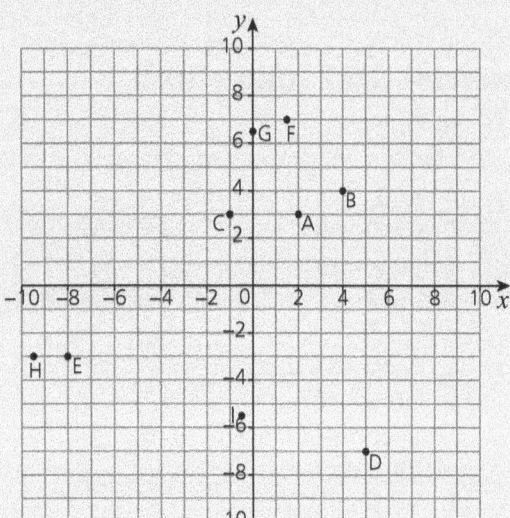

2

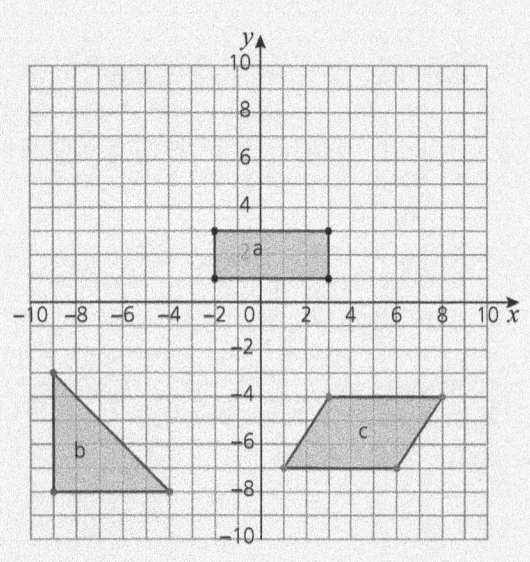

  a Rectangle

  b Right-angled triangle

  c Parallelogram

3 a

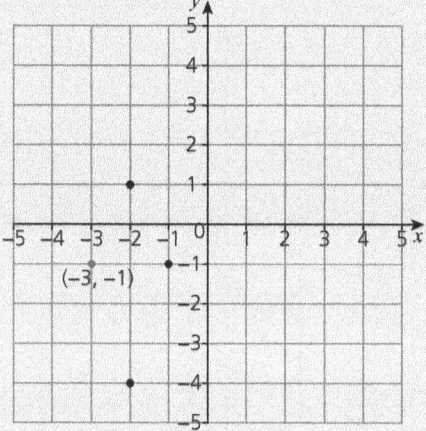

  b Any of the following points or along the perpendicular bisector of the two points given (not shown)

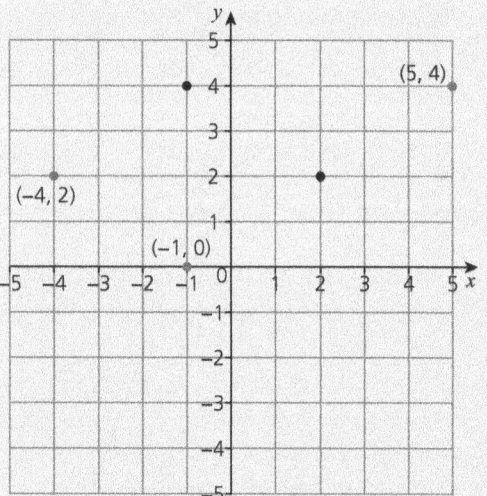

c Any of the following points or anywhere along the lines shown

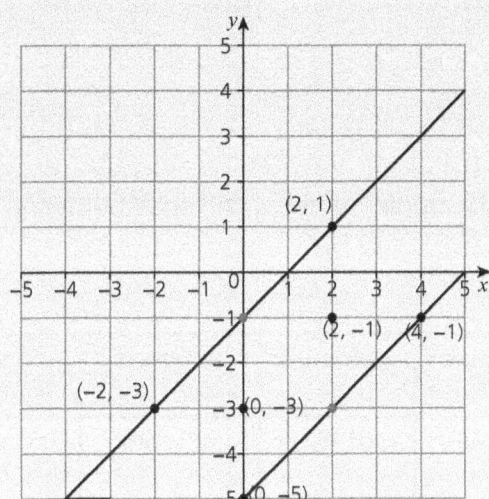

Exercise 25.3 (page 80)

1

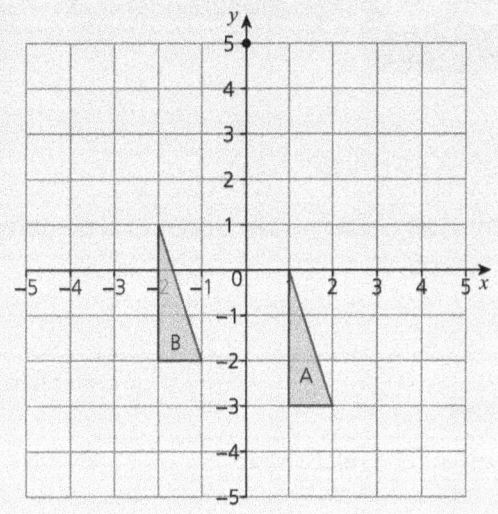

2 (−2, 2) (−1, 4) (1, 2) (2, 4)

3 a (−3, 2) (−3, 6) (0, 2)
  b 5 units to the right and 1 unit up.
  c Right-angled triangles. Two coordinates have equivalent x-values, and two coordinates have equivalent y-values.

4

26 Squares, square roots, cubes and cube roots

Prior knowledge

Students will already know about and be able to identify square numbers and cube numbers.

Objectives overview

Learning objective	Objective code	Student's Book pages	Workbook pages	Teacher's Guide pages	Online resources
Understand the relationship between squares and corresponding square roots, cubes and corresponding cube roots.	7Ni.06	205–210	81–82	120–123	Flashcards Unit 26 Knowledge test Unit 26

Background information

In this unit, students will revise what a square number is and learn how to write it in index form. They will learn corresponding square roots and how to enter and determine these using a calculator. Learning is extended to include cube numbers and cube roots in the same manner.

Students practise their understanding by completing questions on this concept in real-life scenarios.

Terminology

Students should learn and understand the language and symbols associated with the objectives of this unit: square number, square root, cube number, cube root, indices.

Lesson ideas

Square roots are the reverse of squaring and examples show how to calculate the square root of some decimals. Teachers should talk about who Fibonacci was and give examples of the sequence in nature. Cubes and cube roots are taught in a similar manner to squares.

Starter activity

 Who turned out the light?

Mr Light tests lightbulbs by flicking the switch of each bulb to make sure they turn ON or OFF.

When Mr Light flicks the switch of a lightbulb then he switches a bulb that is
- ON to OFF
- and OFF to ON.

Mr Light has 30 bulbs to test; each bulb starts by being OFF.

Mr Light then switches every lightbulb ON.

Mr Light then flicks the switch on every 2nd lightbulb – so now some bulbs are ON and some are OFF, see the table below.

Next, Mr Light flicks the switch on every 3rd lightbulb.

Then Mr Light flicks the switch on every 4th lightbulb and so on.

Mr Light continues in this way until he tests every 30th bulb.

1 Which lightbulbs are ON when Mr Light finishes his test?
  Use the table below to help you keep track! The table shows Mr Light's first three rounds.

Bulb number	1	2	3	4	5	6	7	8	9	10
On/off?	off	off	off	off	off	off	off	off	off	off
	on	on	on	on	on	on	on	on	on	on
		off		off		off		off		off

Bulb number	11	12	13	14	15	16	17	18	19	20
On/off?	off	off	off	off	off	off	off	off	off	off
	on	on	on	on	on	on	on	on	on	on
		off		off		off		off		off

Bulb number	21	22	23	24	25	26	27	28	29	30
On/off?	off	off	off	off	off	off	off	off	off	off
	on	on	on	on	on	on	on	on	on	on
		off		off		off		off		off

2 Mr Light now tests 100 lightbulbs in the same way.
  Which lightbulbs are ON when Mr Light finishes his test?
3 What is special about the lightbulbs which stay ON?
  Why do some lightbulbs end up staying on?

Answers

1 Lightbulbs with numbers 1, 4, 9, 16 and 25 remain ON.
2 Lightbulbs with numbers 1, 4, 9, 16, 25, 36, 49, 64, 81 and 100 remain ON.
3 Square numbers remain ON as they have an odd number of factors, e.g. the factors of 25 are 1, 5 and 25 so this bulb goes ON, OFF, ON. All other numbers have an even number of factors, e.g. the factors of 10 are 1, 2, 5 and 10 so this bulb goes ON, OFF, ON, OFF.

TWM activity notes

Exercise 26.1 is explained here as an exemplar of Thinking and Working Mathematically (TWM) as detailed in the Introduction to the *Teacher's Guide* page xi. Students have been introduced to square numbers and square roots.

Q4 A tiler has 100 square tiles for tiling two square panels in a bathroom.

a Explain why the panels cannot be of the same size.
b Each tile is 10 × 10 cm. What are the dimensions of the two square panels if all 100 tiles are used?

Students are being asked to reflect on what is a square number and deduce that as 50 is not a square number the panels cannot be the same size. Part (b) is a multi-step problem in that students are expected to know the square numbers up to 100 and identify which pair sum to 100 and then work out the side length of each square. The 'Let's talk' question asks students to see if other square numbers can be added to produce a square answer. This is an open-ended investigation, which some pupils will see as a challenge.

TWM characteristics: Specialising Generalising Conjecturing Convincing Characterising Classifying

This question if written in a standard way could have been presented as follows:

Q4 A tiler has 100 square tiles for tiling two square panels in a bathroom.

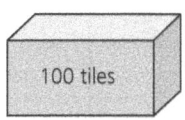  

a Write a list of all the square numbers from 1 to 100.
b If the square panels are of different sizes, how many tiles are used for each panel, assuming all 100 tiles are used?
c Each tile is 10 × 10 cm. What are the dimensions of each of the two square panels?

Students are still expected to know their square numbers from 1 to 100. However, given the list, the task of identifying two which add to 100 is made significantly easier.

Part (c) is a two-step problem, in that students need to calculate the number of tiles along each side first and then calculate its length. This part could be considered as a fairly straightforward TWM type question.

Extension activity

 ## Chess

This extension activity could be used at any point during teaching and learning in this unit.

How many squares are there on a chessboard?

Hint: The answer is more than 64.

Work systematically so you don't miss any!

Answers

Size of square	Number
1 by 1	64
2 by 2	49
3 by 3	36
4 by 4	25
5 by 5	16
6 by 6	9
7 by 7	4
8 by 8	1
Total	**204**

Student's Book answers

Exercise 26.1 (page 207)

1 a, b i 5   ii 3
          iii 11   iv 13

2 a 20
   b i 14   ii 20

3 a $\dfrac{1}{3}$ b $\dfrac{1}{7}$ c $\dfrac{2}{3}$ d $\dfrac{3}{10}$

   e $\dfrac{5}{6}$ f $\dfrac{7}{9}$ g 0.1   h 0.3

4 a They cannot be the same size because
      50 is not a square number.
   b 80×80cm and 60×60cm

5 a 273
   b Because 273 is not a square number.

Exercise 26.2 (page 210)

1 a 64   b 216
   c 1000   d 729

2 a 1331
   b 8000
   c 15.625
   d 238.328

3 a 500 is not a cube number as
      $\sqrt[3]{500} = 7.937$ (3 d.p.)
   b 512 as this is 8³

4 a 2   b 5
   c 3   d 10

5 a 4   b 7
   c 6   d 9.28 (2 d.p.)

6 064

7 No, she is one small cube short. One
   2×2×2 cube is equivalent to eight 1×1×1
   cubes.
   Therefore, the box contains the equivalent
   of 124 small cubes. 5×5×5 = 125

Workbook answers

Exercises 26.1–26.2 (page 81)

1 a 49   b 9
   c 27   d 13
   e 4

2 a $\dfrac{1}{2}$ b $\dfrac{4}{10} = \dfrac{2}{5}$

   c $\dfrac{1}{11}$ d 0.7

3 Width = 2m, length = 2m

4 a 144   b 9
   c 216   d 27

5 $661.24

Linear functions

Prior knowledge

Students may need some revision of Unit 24: Introduction to functions. They should know that a function represents a relationship; a situation can be expressed as a linear function. Students should understand that a function has inputs and outputs and that each single input has a single output.

Objectives overview

Learning objective	Objective code	Student's Book pages	Workbook pages	Teacher's Guide pages	Online resources
Use knowledge of coordinate pairs to construct tables of values and plot the graphs of linear functions, where is given explicitly in terms of x ($y = x + c$ or $y = mx$).	7AS.05	211–220	83–85	124–127	Flashcards Unit 27 Knowledge test Unit 27
Recognise straight-line graphs parallel to the x- or y-axis.	7AS.06	211–220	83–85	124–127	

Background information

In this unit, students learn what is meant by 'the equations of a line'. They revise determining a rule between x- and y-coordinates, plot them on a graph and answer questions about additional points that might lie on that line.

Students see examples of linear function graphs used in real life, for example, in currency conversion. They learn about horizontal and vertical lines, how their equations are known and how to determine if graphs might produce parallel lines.

Terminology

The idea of 'an equation of a straight line' needs careful explanation. Everything in the unit follows from this idea.

Lesson ideas

The unit shows a line graph as a pattern of points. There is a detailed explanation and worked examples which form a detailed lesson plan if necessary. The boxes bring common errors to students' attention.

Starter activity

Connect 4

Lily and Jack are playing Connect 4.

The winner is the first person to make a line of four points.

They take turns to mark a point on the grid with • or ×

Lily is • and Jack is ×

They can only mark points with whole number coordinates.

Here is the start of their game.

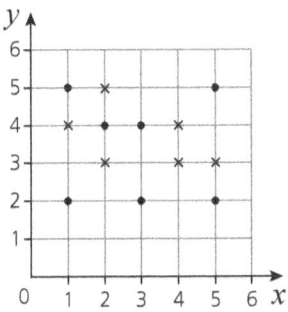

   a It is Jack's turn: where should he go?
   b What are the coordinates of his winning line?
   c What pattern do you notice?
      Complete this sentence:

      'All points on the winning line have a __-coordinate equal to__'

   d Play some games of Connect 4 with a friend.

      Complete parts (b)–(c) above for each winning line.

      What can you say about horizontal or vertical winning lines?

Game 1   **Game 2**

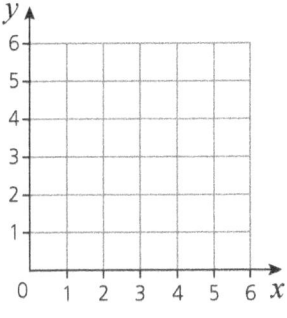 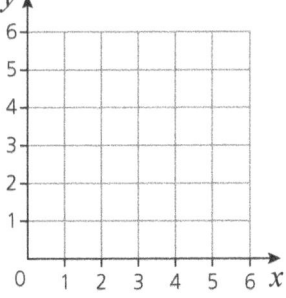

Game 3   **Game 4**

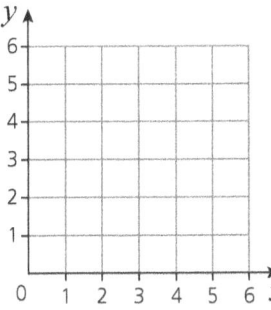 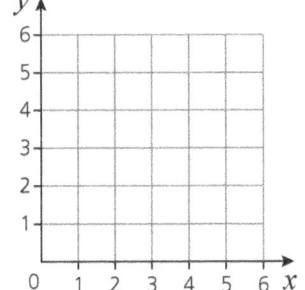

Answers

   a (3, 3)

   b (2, 3), (3, 3) (4, 3) and (5, 3)

   c All points on the winning line have a **y-coordinate** equal to **3**

   d Winning with vertical and horizontal lines should lead to the observation that 'all points on a vertical line have the same x-coordinate' and 'all points on a horizontal line have the same y-coordinate'.

Student's Book answers

Exercise 27.1 (page 215)

1 a i

x	y
0	0
2	1
4	2
6	3
8	4

ii

x	y
0	-2
2	0
4	2
6	4
8	6

   b i The y-coordinates are half the x-coordinates.

     ii The y-coordinates are 2 less than the x-coordinates.

   c i $y = \dfrac{1}{2}x$ **ii** $y = x - 2$

   d $y = x - 2$ because $48 = 50 - 2$

2 a All of them

   b Student's answer

3 a $n = 2.5t$

   b

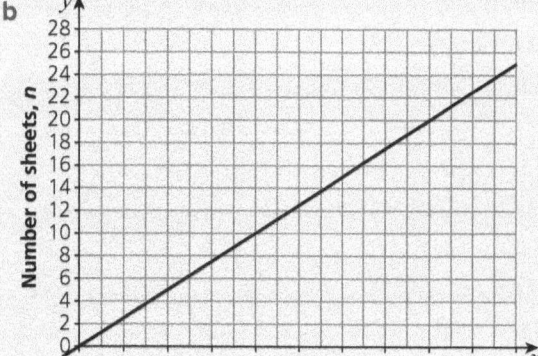

4 a

$y = x + 4$

x	y
0	4
2	6
4	8
6	10
8	12
10	14

$y = \dfrac{3}{2}x$

x	y
0	0
2	3
4	6
6	9
8	12
10	15

   b

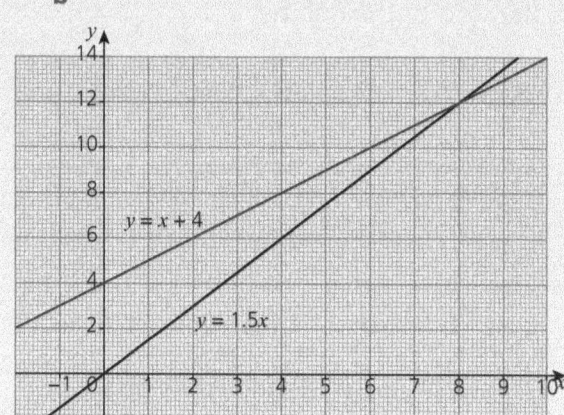

   c The coordinates (8, 12) appear in both tables.

5 a R, S and V lie only on one line. P, T and U lie only on the other line

   b Q

   c $y = 5x$; $y = x + 8$

6 a

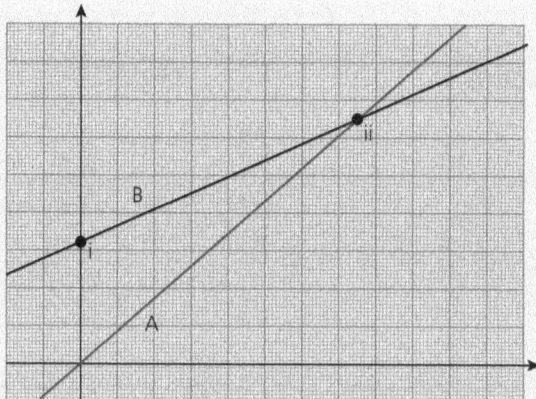

   b i (0, 15) because taxi firm B charges $15 even if no distance has been travelled.

     ii (15, 30) because it is the point of intersection of the two graphs when $y = 2x$ intersects $y = x + 15$.

   c Taxi firm B. If the point of intersection occurs when $x = 15$, this represents the distance where both taxi firms charge the same amount. At $x = 20$, the graph of taxi firm B is below the graph of taxi firm A, meaning it is cheaper.

Exercise 27.2 (page 218)

1 a Students' coordinates leading to $y = 6$
   b Students' coordinates leading to $y = 1$
   c Students' coordinates leading to $x = 6$
   d Students' coordinates leading to $x = 1$
   e Students' coordinates leading to $y = -2$
   f Students' coordinates leading to $y = 0$
   g Students' coordinates leading to $x = -2$
   h Students' coordinates leading to $x = 0$
2 Horizontal lines are a, d, f
   Vertical lines are b, c, e, g
3 a A, B, C and D are on line l_1
      C, E and F are on line l_2

b Point C and the x-coordinate is 8 and the y-coordinate is 6
4 a $y = 4$
   b $x = -4$
   c $y = \dfrac{1}{2}x$
5 a Yes, Three of them. (220, 80), (−100, 200) and (−200, −40) + student's explanation
   b $y = 200$, $y = -120$, $x = 220$ and $x = -200$
   c $320 \times 420 = 134\,400\,\text{m}^2$
6 a $H_{10}: y = 60$
      $V_{15}: x = 120$
   b (240, 120)

 Workbook answers

Exercises 27.1–27.2 (page 83)

1 a

x	0	1	2	3	4
y	2	3	4	5	6

b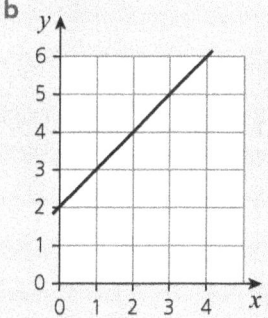

2 $y = 5$, $y = 2x + 1$, $y = 3x - 1$
3 a

x	0	1	2	3	4
y	0	2	4	6	8

b, c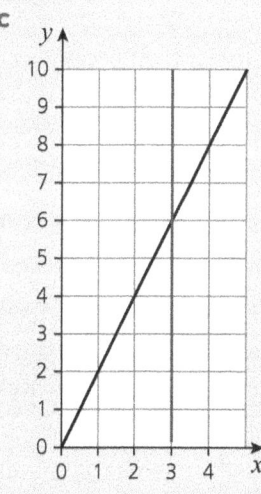

d (3, 6)
4 a $C = 15d$ b $C = 70$
   c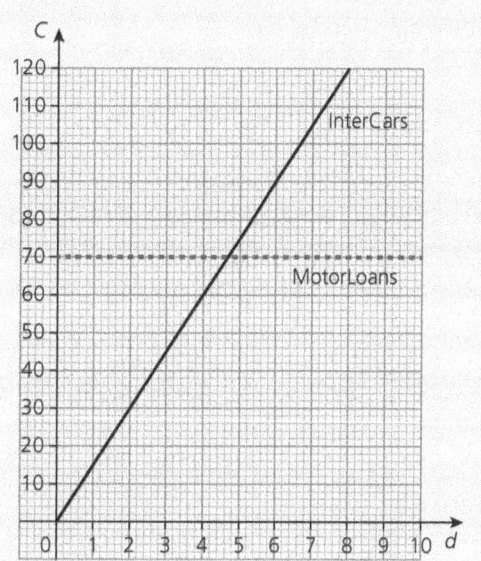
   d 5 days

28 Converting units and scale drawings

Prior knowledge

Students should be confident in the use of metric units of length, mass and capacity.

Objectives overview

Learning objective	Objective code	*Student's Book* pages	*Workbook* pages	*Teacher's Guide* pages	Online resources
Understand the relationships and convert between metric units of area, including hectares (ha), square metres (m²), square centimetres (cm²) and square millimetres (mm²).	7Gg.04	221–229	86–88	128–131	Flashcards Unit 28 Knowledge test Unit 28
Use knowledge of scaling to interpret maps and plans.	7Gp.01	221–229	86–88	128–131	

Background information

In this unit, students revise the metric units of measure for length, mass and capacity and how to convert smaller to larger units, and vice versa. They consider appropriate units to measure different objects and distances. After practising the skills by answering contextual questions, learning progresses to units of area, drawing on students' previous understanding of square numbers.

The unit closes by considering how units of measure might be changed by a scale factor to produce scale drawings.

In Unit 28, students look at the relationship between common units of measurement and how to convert between one and the other. Students also decide on the appropriate units to measure different quantities.

Terminology

Students must know the names of the metric units of measure. They will learn that scale drawings are indicated through the use of ratios which are further explored in Unit 29.

Lesson ideas

The unit begins with the origin of some measures of length and leads on to the metric system. Teachers might want to go into more detail about the origin of the metric system in France. The idea of standard measures can be discussed in groups. An opportunity for discussion of estimated distances can also be useful.

Scale drawing is introduced. A student may have access to a house plan or similar scale drawing. With a few additions from the teacher or from students this unit is a detailed lesson plan.

Starter activity
Measurements wordsearch

Hidden in the grid are 15 words to do with measurements.

Can you find them all?

K	E	H	D	P	L	Z	S	H	E	S	E
I	F	R	N	I	H	Y	T	R	P	S	R
L	J	U	T	S	W	G	T	H	F	A	T
O	W	R	G	E	N	E	Y	B	C	M	I
G	E	R	T	E	M	I	T	N	E	C	L
R	L	V	L	O	R	I	M	D	B	I	I
A	B	K	L	T	N	A	L	A	U	E	L
M	J	I	Y	I	R	N	T	L	R	E	L
D	K	S	T	I	N	U	E	C	I	G	I
M	I	L	L	I	G	R	A	M	E	M	M
E	A	Y	T	I	C	A	P	A	C	H	L
M	E	T	R	E	O	L	T	B	Q	V	D

Answers

KILOMETRE	METRE	CENTIMETRE	MILLIMETRE	HECTARE
TONNE	KILOGRAM	GRAM	MILLIGRAM	LITRE
MILLILITRE	LENGTH	MASS	CAPACITY	UNITS

k	E	H	D	P	L	Z	S	H	E	S	E
I	F	R	N	I	H	Y	T	R	P	S	R
L	J	U	T	S	W	G	T	H	F	A	T
O	W	R	G	E	N	E	Y	B	C	M	I
G	E	R	T	E	M	I	T	N	E	C	L
R	L	V	L	O	R	I	M	D	B	I	I
A	B	K	L	T	N	A	L	A	U	E	L
M	J	I	Y	I	R	N	T	L	R	E	L
D	K	S	T	I	N	U	E	C	I	G	I
M	I	L	L	I	G	R	A	M	E	M	M
E	A	Y	T	I	C	A	P	A	C	H	L
M	E	T	R	E	O	L	T	B	Q	V	D

Student's Book answers

Exercise 28.1 (page 222)

1 a one hundred   b a hundredth
   c one thousand   d a thousandth
   e one thousand   f a thousandth
   g a thousandth   h one thousand
   i a millilitre   j one million

2 a kg   b cm
   c m or cm   d ml
   e tonne   f m
   g litre   h km
   i litre   j cm

3 a–j Students' estimates. Answers may
        vary considerably.

Exercise 28.2 (page 224)

1 a 1m is 100cm, so
      to change from m to cm multiply by 100
      to change from cm to m divide by 100.
   b 1m = 1000mm, so
      to change from m to mm multiply by 1000
      to change from mm to m divide by 1000.
   c 1cm = 10mm, so
      to change from cm to mm multiply by 10
      to change from mm to cm divide by 10.

2 a 40mm   b 62mm
   c 280mm   d 1200mm
   e 880mm   f 3650mm
   g 8mm   h 2.3mm

3 a 2.6m   b 89m
   c 2300m   d 750m
   e 2.5m   f 400m
   g 3800m   h 25000m

4 a 2km   b 26.5km
   c 0.2km   d 0.75km
   e 0.1km   f 0.05km
   g 0.15km   h 0.0756km

5 1kg is 1000g, so
   to change kg to g multiply by 1000 and
   to change g to kg divide by 1000.

6 a 2000kg   b 7200kg
   c 2.8kg   d 0.75kg
   e 450kg   f 3kg
   g 6.5kg   h 7000kg

7 a 2600ml   b 700ml
   c 40ml   d 8ml

8 a 1.5 litres   b 5.28 litres
   c 0.75 litre   d 0.025 litre

9 600ml

10 750ml
11 6550g
12 77km
13 95km
14 138.3 tonnes
15 a 720ml   b 0.53 litre
16 $140.40

Exercise 28.3 (page 226)

1 a 500mm²   b 80000cm²
   c 72000m²   d 6.4m²
   e 56cm²

2 a–c Students' estimates. Answers may
         vary considerably.

3 $587.50

Exercise 28.4 (page 228)

1 a 5m   b 10.5m
   c 4m   d 24m
   e 26.25m

2 a 40cm   b 6cm
   c 15.2cm   d 80cm
   e 40cm

3 a 1:60   b 1:400
   c 1:1500   d 1:5000
   e 1:50000

4 The following diagrams are not drawn
   to the correct scale. Students' diagrams
   should have the dimensions shown.

   a

   b
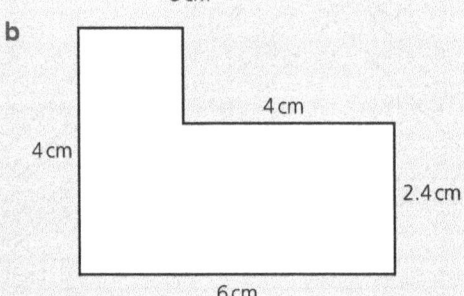

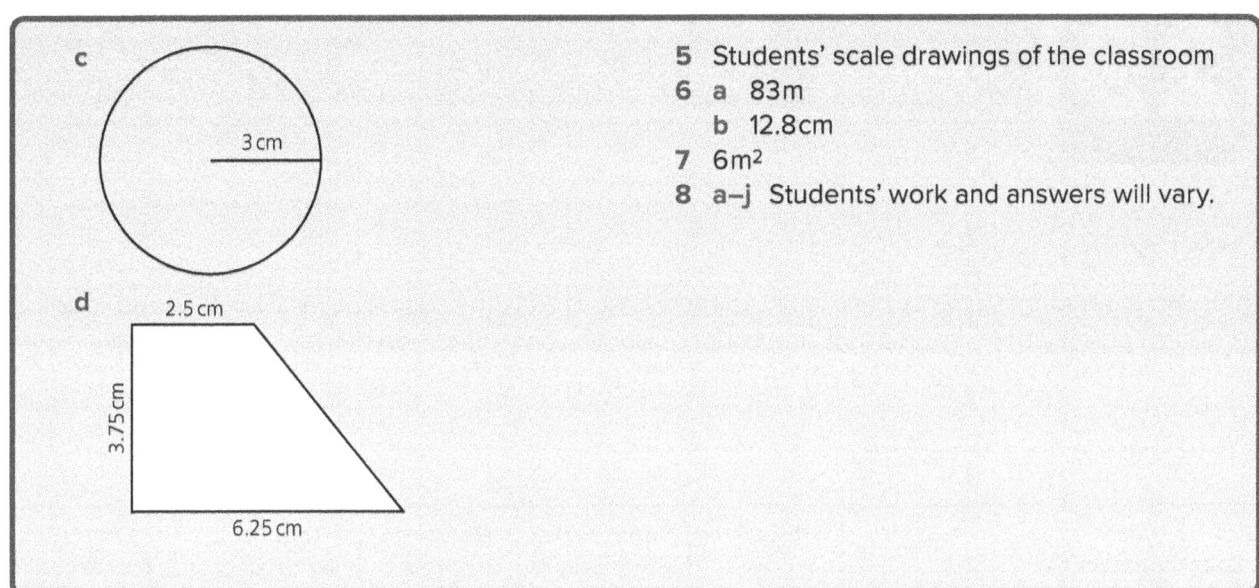

c

3 cm

d

2.5 cm

3.75 cm

6.25 cm

5 Students' scale drawings of the classroom
6 a 83m
  b 12.8cm
7 6m²
8 a–j Students' work and answers will vary.

Workbook answers

Exercises 28.1–28.3 (page 86)

1 a 100   b 10
  c 1000   d 1000
2 a g   b m
  c ml   d tonne
3 a 35   b 8
  c 2000   d 12
4 a 17   b 0.02
  c 2800   d 8.5
5 a 1900   b 500   c 1050
6 a 3.1   b 0.2   c 6.172
7 a 600   b 0.14   c 8
8 1.15 litres or 1150ml
9 900kg or 0.9 tonne

10 30720m or 30.72km
11 $653.75 (523 bags of flour)

Exercise 28.4 (page 87)

1 a 6   b 17   c 16.25
2 a 20   b 3   c 400
3 a 1:50   b 1:200
  c 1:200000
4 6cm by 4cm rectangle
5 a 1420km
  b 2900km

29 Ratio

Prior knowledge

In previous learning, students have had experience working with ratios. They will know that a ratio compares parts with parts and that proportion is about comparing parts to a whole. They will understand direct proportion and will be able to find equivalent ratios.

Objectives overview

Learning objective	Objective code	*Student's Book* pages	*Workbook* pages	*Teacher's Guide* pages	Online resources
Understand and use the unitary method to solve problems involving ratio and direct proportion in a range of contexts.	7Nf.09	230–236	89–91	132–134	Flashcards Unit 29 Knowledge test Unit 29
Use knowledge of equivalence to simplify and compare ratios (same units).	7Nf.10	230–236	89–91	132–134	
Understand how ratios are used to compare quantities to divide an amount into a given ratio with two parts.	7Nf.11	230–236	89–91	132–134	

Background information

In this unit, students use equivalent fractions as a precursor to finding equivalent fractions. Ratios are frequently used in everyday situations, thus in this unit students have the opportunity to study and answer questions using such situations. Students will then use and apply their understanding of ratios to those of direct proportion.

In Unit 29, students explore the use of ratio and how the different ratios of a mixture affect its characteristics.

Terminology

A ratio shows how two or more quantities are related to each other. Ratios, like fractions, can also be simplified. To write a ratio in its simplest form, simply divide by the highest common factor of the numbers involved. The term direct proportion means that two (or more) quantities increase or decrease in the same ratio.

Lesson ideas

The unit compares equivalent fractions and equivalent ratios as revision. The worked example reinforces these ideas. There is further explanation before two exercises. Direct proportion is introduced next with a worked example. The examples, exercises and *Workbook* cover this topic fully.

Students could complete practical activities such as recipe adjustments for larger and smaller servings or paint mixing to make new shades in different quantities to demonstrate this concept.

Starter activity

Logos

Jack and Chloe are designing a logo for the school newspaper.

Here are their designs.

Jack   Chloe

a **For each design find:**
  i how many squares are dark grey, light grey or white
  ii what fraction of each shape is dark grey, light grey or white.
b For each design, copy and complete this statement.
  You should use whole numbers which are as small as possible.
  For every __ black squares there are __ grey squares and __ white squares.
c Jack produces a design that uses only black and white squares.
  For every 1 black square there are 2 white squares.
  What fraction of Jack's design is black?

Answers

a i **Jack's design: 2 dark grey, 6 light grey, 8 white**
   Chloe's design: 14 dark grey, 6 light grey, 16 white

  ii Jack's design: $\frac{2}{16} = \frac{1}{8}$ dark grey, $\frac{6}{16} = \frac{3}{8}$ light grey, $\frac{8}{16} = \frac{1}{2}$ white

   Chloe's design: $\frac{14}{36} = \frac{7}{18}$ dark grey, $\frac{6}{36} = \frac{1}{6}$ light grey, $\frac{16}{36} = \frac{4}{9}$ white

b Jack's design: For every 1 dark grey square there are 3 light grey squares and 4 white squares.
  Chloe's design: For every 7 dark grey squares there are 3 light grey squares and 8 white squares.

c $\frac{1}{3}$

Student's Book answers

Exercise 29.1 (page 231)

1 4:5=8:10=40:50=12:15
2 7:2=14:4=35:10=49:14
3 8:5=80:50=32:20=4:2.5

Exercise 29.2 (page 232)

1 1:24
2 1:14
3 a 4:7
   b 1:75
4 120g
5 1:4:5
6 a i 5:19
      ii 1:3.8
   b 10
   c 455
7 a 2:3:4
   b 480 blue tiles and 640 red tiles
   c 282 yellow tiles and 564 red tiles
8 10 weeks after the start
9 122

Exercise 29.3 (page 234)

1 9 units
2 24 000
3 6000
4 a 1250 minutes (or 20 hours and
      50 minutes)
   b 384 m
5 a 325 g of plain flour, 125 g of sugar, 50 g
      of cocoa powder, 250 g of unsalted
      butter
   b 9 biscuits
   c i No, she has not got enough unsalted
         butter. She would need 700 g.
      ii 39

Exercise 29.4 (page 236)

1 15 boys
2 60kg
3 22 litres
4 36
5 20, 60, 100
6 $x = 48$, $y = 192$, $z = 2$

Workbook answers

Exercises 29.1–29.2 (page 89)

1 a 1:4   b 3:7
   c 6:25   d 1:3
2 a 1:5   b 1:4.5
   c 1:6.5
3 7:8
4 300ml
5 8:7:9
6 a 2:1:1   b 12
   c 24 circular, 12 triangular and 12 square
7 a 0.7m   b 5.6m

Exercise 29.3 (page 90)

1 $105   2 $157.50
3 a 3 hours and 45 minutes
   b 2km
4 No, if you had 1.5kg of sugar from the
   smaller bags it would cost $3.

Exercise 29.4 (page 91)

1 20
2 4.2kg
3 Right-angled triangle as the angles are 30°,
   60° and 90°.

Graphs of rates of change

Prior knowledge
Students will be aware of how to draw line graphs. They will be able to construct tables of values and plot linear functions and will be able to recognise straight line graphs parallel to an axis.

Objectives overview

Learning objective	Objective code	Student's Book pages	Workbook pages	Teacher's Guide pages	Online resources
Read and interpret graphs related to rates of change. Explain why they have a specific shape.	7As.07	237–246	92	135–140	Flashcards Unit 30 Knowledge test Unit 30 End of Section 3 test

Background information
In this unit, students learn about graphs that can represent real-life situations. They learn to interpret graphs without labels from their shapes. They plot their own graphs given measurements and answer questions and draw conclusions about them.

Terminology
The unit starts by introducing a straight-line distance–time graph. The idea of 'constant speed' is discussed.

Lesson ideas
The unit starts by introducing a straight-line distance–time graph. The idea of 'constant speed' is discussed. One of the questions in Exercise 30.1 could be shown on a poster made by students. The unit is in detail and can be used as a lesson plan.

Starter activity

Story graphs

1 Match the graphs with the stories.

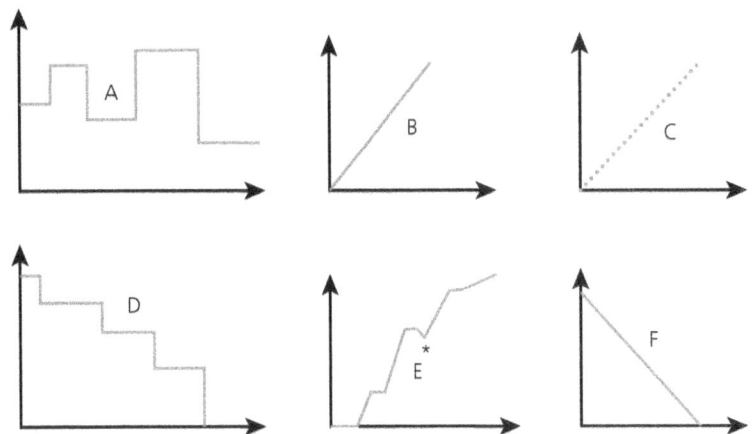

i Amount of birthday money saved against amount of birthday money spent

ii Number of words written for an essay against time after starting essay

iii Cost of petrol against number of litres of petrol bought

iv Amount earned on a delivery round against number of items delivered

v Amount of money in a bank account against time

vi Length of candle against time the candle is burnt for

2 Give a possible reason for the dip marked by * on graph E.

3 Why are the points on graph C not joined up?

4 Make up your own graph and write a story to go with it.

Answers

1 i D   ii E   iii B   iv C   v A   vi F

2 Some of the words in the essay have been erased and then rewritten.

3 The number of items delivered can only be specific values, i.e. it is discrete data, not continuous.

4 Students' own answers.

TWM activity notes

Exercise 30.1 is explained here as an exemplar of Thinking and Working Mathematically (TWM) as detailed in the Introduction to the *Teacher's Guide*, page xi. Students have been introduced to real-life graphs in the context of travel graphs involving distance and time. The significance of a straight-line distance–time graph has also been introduced.

Q2   The distance–time graph below shows the motion of a train over a period of time.

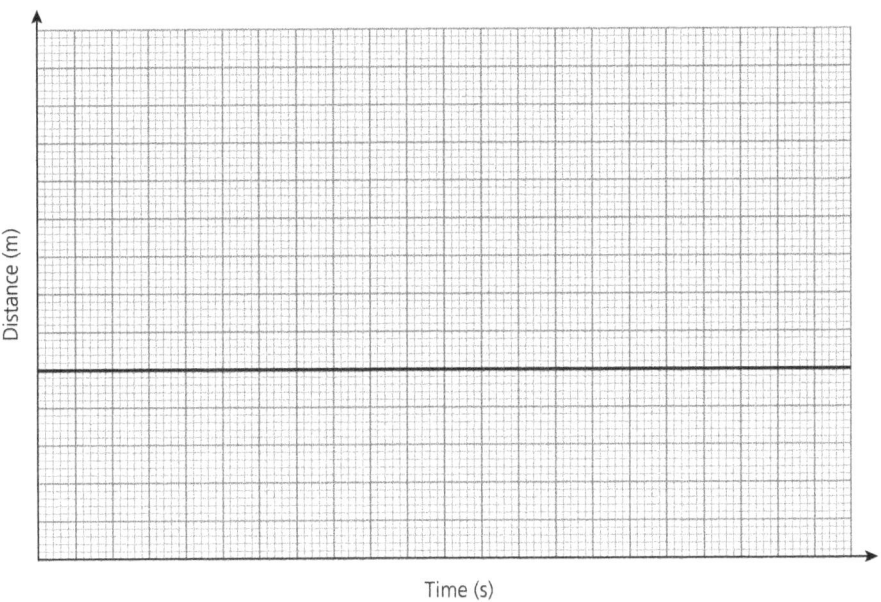

Time (s)

Choose the statement below which best describes the motion of the train. Justify your choice.
i) The train is travelling at a constant speed.
ii) The train has stopped.
iii) The train is travelling on flat ground.
iv) The train is travelling in a straight line.

This question probes the real understanding students have of the meaning of the shape of a distance–time graph. Four scenarios are given, three of which are in fact common misconceptions students have about the shape of the graph. In addition to asking students to justify their choice, the 'Let's talk' box also asks students to explain why they rejected the other three options. Being able to articulate their reasons here is key to demonstrating their level of understanding.

TWM characteristics: Specialising   Convincing   Characterising   Classifying   Critiquing

This question if written in a standard way could have been presented as follows:

Q2   The distance–time graph below shows the motion of a train over a period of time.

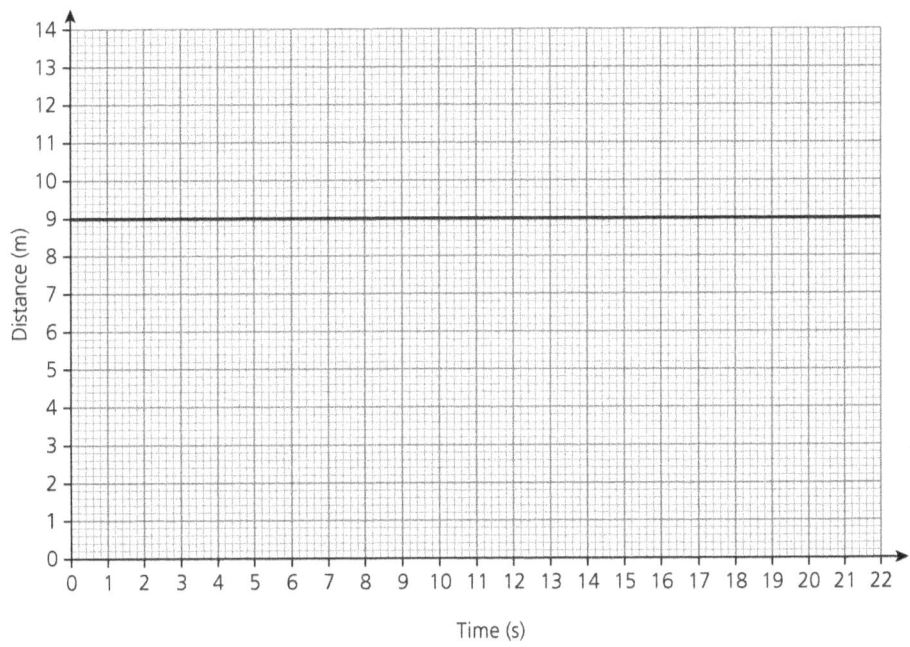

Time (s)

a *How far has the train travelled after 5 seconds?*
b *How far has the train travelled after 20 seconds?*
c *What do you think a horizontal line means on a distance–time graph?*

The conclusion that a horizontal line implies that the train is stationary is the same in both versions. However, the TWM version provides other possibilities which test the depth of a student's understanding. In this case, the fact that the train has not moved has been highlighted in the solutions to parts (a) and (b).

Student's Book answers

Exercise 30.1 (page 239)

1 Graph B as the slope of the graph is least. This means the rate of change of distance with time is the smallest of the three graphs.

2 ii The train has stopped. This is because the graph shows the distance unchanged over time.

3 i The cyclist is travelling at a constant speed
   ii The cyclist is travelling back to a point. As the graph is a straight line it implies a constant speed. As the distance is decreasing with time, the cyclist is returning to a fixed point.

4 a 6.25s   b 40m
   c 8.75s   d 64m
   e 81.25s   f 480m
   g 125s   h 2400m
   i 68.75s   j 28800m (28.8km)

5
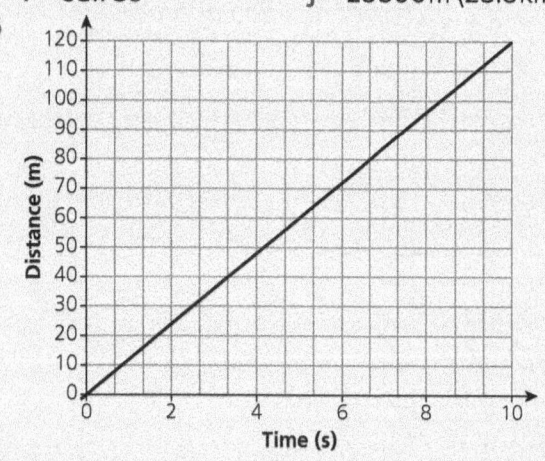

6 a 7a.m.
   b i 50km   ii 50km/h
   c 8a.m.   d 1 hour
   e 50km   f 25km/h
   g 100km/h   h 5 hours
   i 200km   j 40km/h

7 a
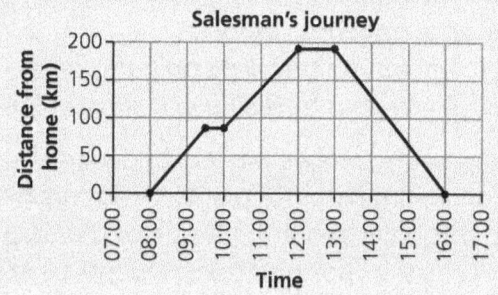

   b $63\frac{1}{3}$ km/h

8 a
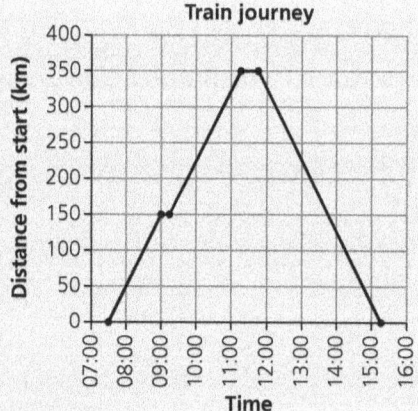

   b 15:15

9 a Students' own lines. Second line must end at 14:00.
   b Students' calculations

Exercise 30.2 (page 245)

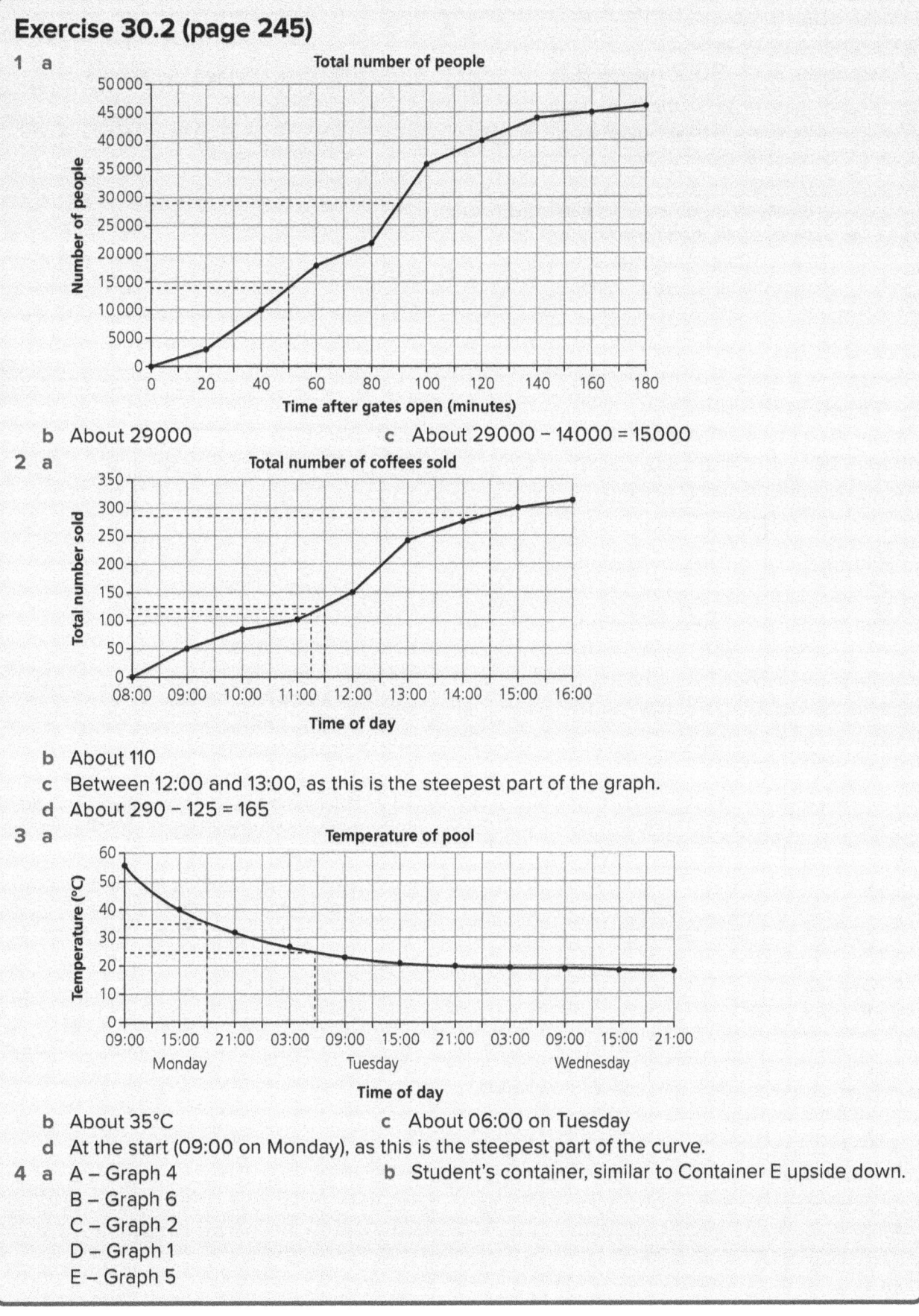

1 a

Total number of people

b About 29000
c About 29000 − 14000 = 15000

2 a

Total number of coffees sold

b About 110
c Between 12:00 and 13:00, as this is the steepest part of the graph.
d About 290 − 125 = 165

3 a

Temperature of pool

b About 35°C
c About 06:00 on Tuesday
d At the start (09:00 on Monday), as this is the steepest part of the curve.

4 a A – Graph 4
B – Graph 6
C – Graph 2
D – Graph 1
E – Graph 5

b Student's container, similar to Container E upside down.

Workbook answers

Exercises 30.1–30.2 (page 92)

1 a 3 or more of any of the following points:
- The coach is travelling at a constant speed for the first 20 minutes for 20 miles.
- The coach is stationary for 8 minutes.
- The coach is then travelling for 1 hour at a constant speed, this speed is lower than the first 20 minutes.
- The coach is then stationary for 12 minutes.
- The coach travels for 20 minutes and travels 40 miles.

   b 20 minutes

   c 10 miles

   d 88 minutes

Section 3 – Review

1 a −2, −1   b $n-8$ c 92

2 a 70%   b 8.3% (2 s.f.)

3 a Accept a rotation of this view

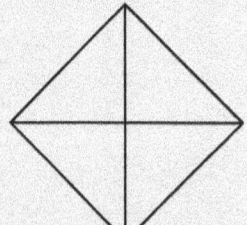

   b

4

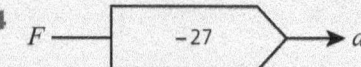

F ──── [−27] ───▶ d

5 C (9, −1)

6 $49 = 6^2 + 3^3 + 2^2$

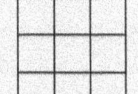

7

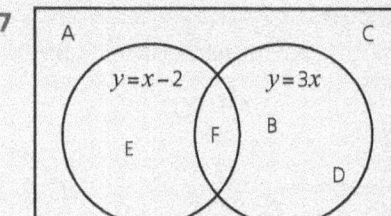

8 32.8mm×58.4mm

9 19 rolls

10 a i Graph Y

   ii It shows the depth decreasing steadily for a bit (when he is filling up the watering can), then remaining the same (when he is walking around watering his plants).

   b i Graph X

   ii The last part of the graph is a horizontal line above the x-axis. This means the depth is unchanged over time, but that there is still water in the water butt.